SUSTAINABLE CITIES - AUTHENTICITY, AMBITION AND DREAM

Edited by **Amjad Almusaed**
and **Asaad Almssad**

Sustainable Cities - Authenticity, Ambition and Dream
http://dx.doi.org/10.5772/intechopen.73410
Edited by Amjad Almusaed and Asaad Almssad

Contributors

Irina Makarova, Ksenia Shubenkova, Eduard Mukhametdinov, Vadim Mavrin, Marco Adonis, Hubert Mukendi, Getachew Assefa, Marwa Dabaieh, Deena El Mahdy, Dalya Maguid, Yupeng Wang, Agnes Schuurmans, Susanne Dyrboel, Patrizia Lombardi, Sara Torabi Moghadam, Stefan Bakker, Amjad Zaki Almusaed

First published in London, United Kingdom, 2019 by IntechOpen
IntechOpen is the global imprint of INTECHOPEN LIMITED, registered in England and Wales, registration number: 11086078, The Shard, 25th floor, 32 London Bridge Street
London, SE19SG – United Kingdom
Printed in Croatia

British Library Cataloguing-in-Publication Data
A catalogue record for this book is available from the British Library

Additional hard copies can be obtained from orders@intechopen.com

Sustainable Cities - Authenticity, Ambition and Dream, Edited by Amjad Almusaed and Asaad Almssad
p. cm.
Print ISBN 978-1-78985-523-4
Online ISBN 978-1-78985-524-1

For EU product safety concerns:
IN TECH d.o.o., Prolaz Marije Krucifikse Kozulić 3, 51000 Rijeka, Croatia,
info@intechopen.com or visit our website at intechopen.com.

Meet the editors

Dr. Amjad Almusaed was born on January 15, 1967. He holds a PhD degree in Architecture (Environmental Design) from "Ion Mincu" University, Bucharest, Romania. He followed a postdoctoral research program in 2004 on sustainable and bioclimatic houses from the School of Architecture in Aarhus, Denmark. Dr. Almusaed has more than 28 years of experience in sustainability in architecture and landscaping with innovative orientation. He has carried out a great deal of research and technical survey work, and has performed several studies in the area mentioned above. He is an active member in many international architectural associations. He has published and edited many papers, articles, research works, and books in different languages.

Asaad Almssad has more than 28 years of experience in the industry as well as teaching and research at Umeå University, Karlstad University, both European and non-European institutions. His research focuses on building structures, materials, sustainable building, and energy efficiency in building systems. His viewpoint of the building and its components is that the orientation of new research tends to move human actions from under the building roof toward energy efficiency and healthy living spaces. He has authored and co-authored more than 30 research papers and many books. He is now employed as a docent at Karlstad University, Karlstad, Sweden.

Contents

Preface

Aristotle defined the city as follows: "A city must be built to provide its inhabitants with security and happiness." Friederich Ratzel defined it as "a sustainable density of human dwellings and people who occupy considerable space and are at the crossroads of a major communications road." Cities are essential to humans. For those who live, work, or visit them, as well as for those who rely on growth, the cities generate for both the city and the country. The emergence of cities on the stage of history was a fiery consequence of the evolution of human civilization. Originally, the city was created to fulfill a commercial product exchange role and a military (security) role. The first urban civilizations in the world emerged in Asia near advanced agriculture systems based on irrigation. The oldest city in the world, Jericho, appeared 7000 years ago in the Jordanian territory. Between the fifth and second millennia BC, flourishing cities developed in Mesopotamia (Ur, Uruk, Babylon), the Mohenjo-Daro Valley, northern China, and Phoenicia (Tyr, Sidon). Starting with the third millennium BC, cities began developing in northern Africa in the Nile Valley (Memphis) and the Mediterranean Sea (Carthage). In Europe, the Minoan civilization created the first urban centers in the second millennium BC (Knossos on the island of Crete). Subsequently, in ancient Greece, there are city-states (Micene, Corinth, Athens). The ancient Greeks also set up numerous colonies in Asia Minor (Ephesus, Miletus) and on the Black Sea coast (Histria, Tomis, Callatis). In the Roman era, cities saw definite progress with paved streets, sewerage and water supplies, public utilities, and central forum markets. Interestingly, many European cities were built on the ruins of the Roman ones (London, Paris, Vienna, Turin). In the Middle Ages, many cities fell. They were fortified with defense walls and ditches. As a consequence, the city is a well-populated regional body with a high degree of concentration, organization, and social, cultural production, formed under certain conditions of space and time. The city can be defined as a spatial, economic, and social formation that cooperates with a multitude of factors with which it is in close interdependence and reciprocity; a body that employs large spaces and has a unique role in polarization or economic gravity factors through production and consumption. We are used to sector-oriented thinking. It is the traditional approach for many local and national initiatives, but in the future, we need to associate traditionally divided sectors. With increased pressure on cities and still fewer resources available, it is imperative that the urban policy of the future be developed in an ever-closer interaction between areas such as transport, business, social, integration, environment, and cultural theory. Cities need to help ensure and improve the quality of life of their inhabitants and business attractiveness by providing sophisticated information and communication technologies in the fields of education, employment, social services, health, and safety. Our cities must be able to adapt to the threat posed by climate change. Properly designed and properly planned urban development can improve the quality of the environment and reduce carbon emissions. Cities need to achieve these results through innovative preventive, mitigation, and adaptation measures that contribute to the development of new industries, and activities that generate a low level of carbon dioxide emissions.

It is difficult to determine precisely the date of birth of the sustainability concept in concern to cities. Sustainability in the urban domain appears, at minimum apparently, to have emerged out of the two unified oil crises in the late 1960s and early 1970s: the first of these crises was ecological, the second urban. The environmental crisis was a product of the culmination of large- and small-scale environmental damage that had been brought by rapid industrialization. Sustainable urban development takes place in many cities as densification and transformation. It's the city's goal, first and foremost, to develop within the existing limits so that no larger new areas are included. Many municipalities are working strategically to fortify city centers and to transform function-based commercial and port areas, among other things, to renovate or expand already built structures and regions as well as recycle closed-down commercial properties, port areas, and other functional emptied buildings/land for new purposes. The public functions for the city's life, and environmentally, a large number of urban functions today can be well integrated into the urban environment. To support this, there is a need for methods and tools to help municipalities and others to be inspired to plan for existing and future businesses, which need to be in interaction with the city. The properties of the state, regions, and municipalities can also contribute to sustainable urban development by, among other things, working with location strategies that combine different institutions in, for example, multipurpose houses, so that operating consumption can be reduced. Building flexibility can be increased, and proximity to infrastructure can be secured. The concept of densification is that one always tries to think of further and future needs; it goes together with planning and development of healthy cities.

In this book, many leading experts, including urban planners and academics, have collectively expounded and shared their concerns and strategies on the new vision of sustainable city movement in our world today. It will be a "must-read" book for a broad market , including city decision makers, academics and researchers, the public, private sector professionals such as planners, architects, engineers, landscape designers, geologists, economists, etc.

The book is divided into four parts and nine chapters:

Section I is entitled "Introduction to Sustainable Cities Concept," and contains one chapter entitled "Introductory Chapter: Overview on Sustainable Cities Theory and Practices," which discusses sustainability in cities in concept and practice.

Section II is entitled "Energy and Environmental Analysis of Sustainable Cities Models." This include four chapters. It expresses the effect of environment and energy on city configuration and functions. The second chapter, "Towards Adaptive Design Strategies for Zero-Carbon Eco-Cities in Egypt attempts to describe the hidden potentials by way of analyzing successful private initiatives for existing eco-communities in Egypt, where an analytical case study method for tackling different aspects such as renewable energy, permaculture, eco-sanitation, solid waste management, vernacular architecture, green transportation, and green economy was adopted. The chapter contributes by critically analyzing such attempts and concludes with design recommendations and strategies on how to reach an environmentally enriched, healthier, resilient, and socially rewarding zero carbon city, running on its own locally available resources. The third chapter is "Buildings in Urban Regeneration." In this chapter, the authors try to elucidate that the built environment process is a key element in urban living, where by taking a holistic approach, buildings have the potential to be part of the solution to today's and tomorrow's challenges through the creation of sustainable cities. The fourth chapter of the book is "Urban Renovation and the Simulation Evaluation of Urban Climate Change in Residential and Commercial Districts: A Case of Xi'an, China". Urban heat island is the most important issue in this chapter, which requires the monitoring and evaluation of outdoor thermal comfort in cities worldwide, where four microscaled residential and three commercial districts in Xi'an city represent the typical urban typology of residential and commercial districts that were developed during different historical periods, and use the urban simulation system scSTREAM to evaluate the impact of urban renovation types on urban climate change. The fifth chapter is "Multi-criteria Spatial Decision Support System for Urban Energy Planning: An Interdisciplinary Integrated Methodological Approach." This chapter provides an interdisciplinary integrated methodological framework, which provides guidance to develop a multicriteria spatial decision support system that supports decision-making processes for urban energy planning purposes. The chapter helps in defining and evaluating energy-saving scenarios, taking into account the participation of stakeholders in an interactive way.

Section III is entitled "The Role of Transport in a Sustainable City." This part includes two chapters that begin with Chapter 6 "Electric Two-Wheelers, Sustainable Mobility and the City.". The chapter explores the role electric two-wheelers (including pedelecs, e-mopeds, and e-scooters) can play in urban vehicle ecosystems using the sustainable mobility paradigm. Compared to traditional transport planning, this paradigm has a stronger focus on aspects such as accessibility, people, streets as a space, city livability, as well as environmental impacts. The analysis is based on existing literature in the academic and policy realm, and a comparison with other transport modes, including motorcycles, bicycles, public transport, and cars. The seventh chapter is "Influence of the Motor Transport on Sustainable Development of Smart Cities." This chapter focuses on the benefits of using smart transport in all fields of life as well as intellectualization of the decision-making process.

Section IV is entitled "The Influence of Social and Economic Factors in Urban Space Conception" and contains two chapters. Chapter 8 is "Life Cycle Insights for Creating Sustainable Cities" and focuses on the full life cycle of materials and energy flows as well as their uses in cities. The chapter concludes with recommendations on best practices that potentially leverage life cycle assessment results. It also covers the merits of employing the social life cycle perspective together with the environmental life cycle and economic life cycle in a life cycle sustainability assessment framework that seeks to define the triple bottom line space of lower unsustainability conditions. Chapter 9 is "Smart Homes and Sustainable Cities: The Design of a Low-Cost Solution for Comprehensive Home Automation," and presents a broad overview of the design and development of a web-enabled smart home solution. Web development and control systems together form the backbone of modern home automation technologies such as the Internet of Things and embedded systems.

This book offers a variety of materials on sustainable cities and eco-cities, and will prove useful to professionals in the field of urban planning and design. We want to express our special thanks to all the authors who contributed to the quality, range, diversity, and richness of this publication with their chapters. A special thanks also go to the kind Mr. Julian Virag, InTech's Publishing Process Manager, for his assistance and efficiency in the management process of this book and his cooperation at various stages of book publication. Last but not least, we wish to acknowledge the superb assistance of the staff at InTechOpen Publisher in the preparation and coordination of this book.

Amjad Almusaed
Jonkoping University, Sweden

Asaad Almssad
Karlstad University, Sweden

Section 1

Introduction to Sustainable Cities Concept

Chapter 1

Introductory Chapter: Overview of Sustainable Cities, Theory and Practices

Amjad Almusaed and Asaad Almssad

Additional information is available at the end of the chapter

http://dx.doi.org/10.5772/intechopen.82632

1. Introduction

Human settlements are the result of the dynamic adaptation of the human community operating in a given territory in the conditions of social, economic and historical relations. The areas on which human settlements are located are distinguished by the components of the physical-geographic structure, by the diversity and by the potential natural conditions, as well as by the economic and social factors in which the human settlements appear and develop [1]. Human settlements represent the totality of human communities, villages and towns, regardless of their position, size and functions. Human settlement can be considered a geographic landscape integrated with the natural and social conditions necessary for the existence of housing, work and equipment (power supply, water, transport, communications, sanitation, etc.). Human settlement is a body of land known to be a regular form of property with a hearth on which communal and territorial attributes develop [2]. The term "locality" defines a human, rural or urban settlement, delimited according to the number of inhabitants, the nature of the built-up area, the degree of the social endowment, the technical-public amenities, the function, etc. Hence, human settlements or human habitat refers to some components such as population, construction, markets, streets, industrial platforms, recreation and recreation areas. Human settlements support the unity of natural, social, material, spiritual, cultural and organizational factors, including housing, labor, energy supply, communications, water, sanitation, services, social security, administration systems, cultural facilities, recreation, etc. [1]. The village is the oldest form of human habitation that presents ethnographic, historical, economic, social or urban characteristics. A village is a group of houses and people who are leaving their means of existence of a determined social space.

Defining the sustainable city, the word sustainability has come into such common usage that it sometimes seems ubiquitous. At the outset, this leads to the need to answer two principal questions with regard to the movement of sustainable cities. First, what is a sustainable city? And secondly, why is it important that cities become sustainable? In answering these questions, it is useful to draw a distinction between sustainability and sustainable development. Sustainability in its broadest sense is the capacity of natural systems to endure and to remain diverse and productive over time. Sustainable development is the practice of humans arriving at a level of economic and social development that does not inevitably alter ecological balance [3]. Many settlements were fair, with trade, crafts and administrative functions. In the renaissance era, when the bourgeoisie grew, the cities experienced a period of flowering (Florence, Venice and Rome). In the modern and contemporary era, besides the European and Asian cities, colonized colonists have developed in the Americas (New York, Quebec, Bogota). A major urban transformation accompanied the industrial revolution. Between 1800 and 1990, global population grew almost six times, while urban population multiplied more than 120 times. The human settlement represented by urban settlements has two basic components:

- Territorial components that are superimposed on the built perimeter, this being what we call the hearth, delimit the space for the living quarters; sometimes overlaps with the city.
- The socioeconomic and environmental component refers to the population and to the work place, which are closely linked to the city's territory.

The commune is a form of territorial administrative organization that includes exclusively rural settlements, namely villages, whose common point is represented by the unitary, social, cultural or ethnographic character. One or more villages may enter it. The city was defined by F. Ratzel by the existence of three elements: professional activity, concentration of dwellings and number of inhabitants. "The city, as an element of the landscape, is a body linked to the geographical space, inside of which it fulfills a precise function: it concentrates, transforms and redistributes material and spiritual assets." [4]. It is necessary to consider the marked urbanization that draws the population and the growth of the cities. About half of all people live today in the big provincial cities. The majority of Western cities possess unique cultural and architectural qualities, strong social inclusion forces and exceptional economic development opportunities. Cities are centers of knowledge and sources of growth and innovation. Still, it faces problems linked to demographic and social inequality as well as social exclusion of specific population groups including a lack of affordable housing and environmental issues. In the long run, cities will not be able to fulfill their function of engines of social progress and economic growth as described in the Lisbon Strategy, if we fail to maintain social balance in cities ensuring cultural diversity and setting high-quality standards in the areas of urban design, architecture and environment. It is necessary to increase the need for integrated strategies and coordinated actions by all individuals and institutions involved in urban development that can have a more general impact than only cities that are taken individually. There must be a responsibility for the future of our cities at every level of government—local, regional and international level. To make this multilevel governance genuinely useful, it will be necessary to improve the coordination of the various sectoral policy areas and develop a new sense of responsibility for integrated urban development. Also, it is necessary to make sure that those working on developing these policies at all levels learn the skills and knowledge that they need to build sustainable communities in cities. In modern definition, the city should be conceived not only as a packing of houses and people with a certain physiognomy, with an economic and social life limited to a narrow frame, but especially as a complex organism with much wider functions whose area of manifestation goes far beyond that of the city built in other parts.

2. Healthy cities and energy efficiency

According to UN programs, we will be 10–12 billion people around the world by 2050. Agenda 21 is an action program adopted at the United Nations Environment and Development Conference in Rio de Janeiro in 1992. The program describes how efforts to counter environmental degradation, poverty and lack of democracy should be developed in order for our societies to achieve sustainable development. Agenda 21 notes that the human way to extract and use the energy inside is permanently sustainable and identifies two ways to change the energy system: energy efficiency and renewable energy sources. In order for everyone to have a reasonable material and healthy standard of living, natural resources must be utilized efficiently. Economic growth is high in Asia, for example, concurrently the population increases. This leads to more cars, more mobile phone and tablets, as well as white goods such as refrigerators, higher power consumption and so on. Pressure on finite resources like oil, coal and gas becomes large. But the most difficult issue is the pressure on biological resources: forests, arable land, fisheries, wetlands and mangrove swamps [5]. The existence of economic activities and investments on the one hand is closely linked to the existence of high-quality urban structures, a properly constructed environment and a modern and efficient infrastructure. For this reason, it is necessary to improve the existing building opportunities in disadvantaged areas, in terms of design, physical conditions and efficiency of energy use. Adaptation of housing standards for new and existing buildings has the greatest potential for increasing energy efficiency within the EU and thus combating climate change. In order to increase the sustainability of investments in improving the physical environment, they must be included in a long-term development strategy that includes, among others, a sustainable public and private investment program. The increasing population growth puts pressure on, among other things, urban areas, resources, housing prices and public services. Additionally, increased traffic volume results in poorer air quality and congestion-related problems. At the same time, there is also an increasing trend towards social divisions between populations in the larger city. The new challenge for authorities and city planners is to transform metropolitan areas into eco-areas, in a form "sustainable cities," which is able to survive only with green energy and reduce pollution as much as possible. More than half of the world's population lives in cities today, according to the United Nations. Given the unprecedented levels of urban migration in recent decades, in conjunction with climate change, it is necessary to recognize that human livelihood in a significant urban agglomeration, had a big impact on nature. For cities to find a place in the green future of the Planet, they have to turn into clean entities. There are already cities in the world where the future is here. Citizens of the capital of Iceland, Reykjavik and those in Vancouver, Canada, consume energy supplied almost exclusively from renewable sources. Vancouver is somewhat blessed geographically—with mountains, rivers, oceans and valleys—but the citizens have always tried to help them and make the most of what nature has provided [6]. Considering the complexity of the challenges, there is a need for an overall urban policy framework for how they are addressed and the potentials being exploited. Thinking of social conditions, environment and economy together improve the likelihood of implementing coherent solutions. Sustainable settlements must be economically viable while incorporating climate adaptation, energy and resource efficiency, environment, architectural quality and social security. Economy and the environment can thus generate added value when combined with overall considerations. Urbanization can contribute to a more sustainable society, including linking different city functions into urban development strategies. For example, proximity to public transport can reduce resource and energy consumption when planning accommodation, jobs and shopping opportunities so that busses, trains, walking and cycling are preferred. World cities occupy only 4% of the land area, yet they are home to more than half the world's population. Since 2009, the number of inhabitants in these cities has increased by 7.6%. The development of sustainable cities is one of the most significant global challenges right now. The cities face a large number of social, environmental and economic problems that require conversion. As population flow brings new demands to physical structures, to residents and the management of cities, the need arises for thinking across professions and sectors. This conversion can be done through public-private partnerships where administrations and companies benefit from each other's expertise and experience. Today, each year, humanity uses energy equivalent of only a few thousands of a percent of the solar energy that reaches the Earth's surface at the same time. The types of energy used are just over 90% fossil fuels, according to some analysts.

3. Innovative sustainable cities and urban policy

A sustainable framework for urban policy should not only focus on the future but also have a strong contemporary focus, as it will quickly become impractical in everyday life. The vision for a new framework for sustainable urban policy can, therefore, deal with methods that make sustainable solutions an attractive and beneficial alternative for all. Sustainable solutions are intended to be based on Dane's daily needs. For example, we are not necessarily exclusively cycling because it is environmentally friendly and healthy, but perhaps because it is easy, fast, cheap and accessible. This principle can be transferred to sustainable urban development. Everyday life cannot be more cumbersome by a sustainable change, so it is likely that the broad favorite anchorage will be left out. In many occasions, we used several new terms in sustainable city field. There have been confusions in the use of some terms such as climate smart, sustainable development, carbon-neutral and not least of all the environmental classifications of goods and services which appear at regular intervals. It is not surprising if different actors may feel uncertain about what actually makes the greatest benefit for the environment. Initially, it may be useful to define the concept of sustainable development, because it has had such a significant impact on sustainable urban development becoming a political and even commercial focus area. In the literature, there are many different definitions of what sustainable development is. A brief description is to maintain a positive social change process [7]. Thus, it is the whole society to participate in a process in which the aim is that people's needs must be satisfied, without spending too much on the Earth's resources. However, it is necessary to recognize the limits, conversion and the real meaning of sustainable cities, where the conversion to sustainable cities is therefore both an individual and collective project that can support the city's communities. It requires shared ownership for all, and in the process of initiation, it is important that the city's users get involved in the thinking of the solutions. The government's sustainable urban policy is primarily about social and biophilic theory. In cities that are growing rapidly, poverty also increases. The problem with the cities is that it is difficult to develop infrastructure and services in line with population growth. This means that the living conditions and environment deteriorate for lots of people. Many of the developing countries will continue to grow rapidly. But all cities do not grow, and growth in many megacities (with more than 10 million inhabitants) has slowed down in recent decades, for example, in Latin America's largest cities. The same applies to many Asian cities, such as Calcutta, India. Most of the largest cities are in a handful of states: China, India, Brazil, Indonesia, Mexico, Egypt and Pakistan. In more than 50 nations, there is no city that has reached half a million inhabitants. Third World cities, including some of the largest, are not very populous if counting people per hectare [5]. There will be water shortages, environmental degradation, traffic congestion, slums, crime and abuse. Urbanization also has a positive side, partly because it is strongly linked to economic growth. A sense of social and biophilic city can also have a positive effect on growth and employment as well as the framework for the good life.

3.1. Sustainability and social responsibility (a socially sustainable city)

Social sustainability revolves around the human factor as a prerequisite for a sustainable city and a sustainable society. The government's sustainability strategy thus denotes the notion that everyone should participate in the social development and have equal opportunities, regardless of the background. A focused and sustained social sustainability work helps to ensure diversity, democracy and equality in cities. A sustainable city is socially linked to the fact that there are democratic spaces where people can meet regardless of social, economic and cultural backgrounds and provide opportunities for deployment and accessibility for all the citizens of the city. When a city has many offers for both everyday life and special occasions, it becomes more vibrant and attractive—this can also help to increase the quality of life for urban citizens. A socially sustainable city also opens the possibility for citizens' health to be supported in the form of urban spaces designed for physical activity and intercourse. The population's living patterns have also changed. Now, we live in the age of individualism. More and more people spend periods in their adult lives where they live alone. This affects the demand for housing and transport. At the same time, old-day collective solutions are no longer as relevant as they once were, either in the form of commercial offers or in the form of totally disorganized and spontaneous practice of sports. It is important that the cities also allow for individual sports and physical activity. This can help residents improve health and also the cohesion between people who might not otherwise meet each other. At the same time, a city and the residential area must be experienced attractive enough for a healthy and natural exchange with the surroundings and that different people want to visit, live and settle there. Urban development is crucial for communities—both the large community at the urban level and the closer communities around the residential areas. An essential element of social sustainability is that there is a varied range of housing types—types of homes, sizes, location and different prices, including a housing that gives everyone, regardless of financial ability, the opportunity to have a reasonable framework for everyday life. The socially sustainable city also ensures that people with social problems get the necessary support to maintain the housing.

The general sector helps to ensure socially sustainable cities that support social and economic balance. In a broad sense, the legitimate resident democracy can be part of a sustainability concept. It is important to ensure social diversity in general housing so that citizens with economic, social or integration challenges are not concentrated in particular areas or neighborhoods. For a number of years, efforts have been made to ensure a better balance in the so-called vulnerable residential areas where such imbalance is a reality. In addition to social efforts, a socially sustainable city is being pursued by building new attractive general housing on a smaller scale and in neighborhoods with mixed ownership and housing forms, which are physically integrated in the surrounding city. This should help to ensure a better social balance in the residential areas, thus avoiding further segregation and division. The office often functions as a gathering place for homeless people, addicts and people with mental difficulties. Social sustainability means creating the necessary framework in the city for vulnerable groups: both in the urban area itself or the form of homes for those groups. The socially sustainable city prevents the exclusion of the city's communities and ensures that the basis for social action for vulnerable citizens is present. Relationships between the physical and social environment are central to social sustainability. Citizens' involvement in the development of cities is essential for social sustainability, involving city citizens and residents in promoting shared ownership and sense of responsibility, which the municipalities are working extensively in connection with area renewal. Citizens' involvement is also practiced in many other contexts, such as center planes, port plans or other. It is therefore necessary to continue developing methods of inclusion, including in terms of activating and engaging citizens who might not otherwise be involved—including using digital engagement platforms. Citizen involvement should always take place on the citizen's grounds irrespective of cultural, economic and social circumstances.

3.2. An environmentally sustainable city (biophilic city) and climate change phenomenon

A logically interpretation of a Biophilic model in sustainable cities is oriented to maintain a clean environment, access to nature and climate change management, resource shortages and pollution, which requires the conversion of our cities and homes. This applies, for example, to promoting sustainable modes of transport, reducing energy consumption in buildings, establishing new forms of energy supply and other ways of managing resource streams. A large number of environmental problems and challenges are concentrated in the cities, and at the same time, the solutions must be found. Urban development, housing enrichment and construction will aim at a wide range of solutions that together will be the green and blue city, where pollution and effects thereof are minimized and where nature and water are incorporated into urban solutions and become more visible in the cityscape than today. The cities have a great potential to develop and implement innovative solutions and to reduce energy and resource consumption. For example, energy consumption and congestion can be reduced by planning a compact urban development with good public transport, making it more attractive to choose public transport, bicycle and corridors. The consequences of climate change can be felt even more clearly in the future—flooded basements and congested sewers are partly results of climate change in combination with the vast fortified areas that prevent rainwater from leaching into the soil. There is a need to think across and to reap the benefits of incorporating social elements and economics into, for example, climate adaptation initiatives. If climate change adaptation is combined with the city's space and life, new opportunities are opened to incorporate environmental solutions in the city. The climate problem is highly relevant for urban development. Reducing carbon emissions by using different policy measures is an important part, in addition to planning new districts in such a way that a changing climate does not provide any unwanted side effects. It is largely the lifestyle that determines whether we live healthy or not. Updating of knowledge about sustainability and climate change is necessary. The basis on which rules and technical specifications were based only 5–10 years ago must be adapted to match this new knowledge. Changes in business and the population's living patterns and the modern understanding of climate change cause all cities to face a series of challenges and opportunities. All cities must work towards greater sustainability. The cities are also different. It will be the challenges and thus the possibilities as well. Today, it is possible to open the ports and islands of the cities to the benefits and enjoyment of the greatest possible extent. The transformation requires careful consideration to make a sensible balance between the many interests of housing, leisure, business and port activities. Otherwise, one kind of shutdown can be quickly replaced by another.

3.2.1. *Environment and climate change*

Climate change and overweight are major threats to health. A major and growing environmental problem being discussed much is climate change. Scientific evidence has shown that the levels of greenhouse gases have increased in the atmosphere and that human fossil fuel combustion has increased the emissions of these gases [8]. This in turn leads to an increase of the annual average temperature on the Earth, which can contribute to droughts, floods and other serious problems against human survival around the world. A city can become a healthier city if it is obvious to go to combat the negative act of climate change, if the city is clean and air quality is good, if the use of hazardous substances both outdoors and indoors can be avoided and if, at the same time, it is easy to find quiet areas where there are green areas and where we can rest, cycle or go for a walk. Increased density must be combined with green, blue and healthy living; however, successfully combining these poses a challenge. This applies to the planning of the cities, town's buildings and to the technical solutions for transport, water, energy and waste. The cities are different—the solutions become different. The starting points for meeting the challenges are not the same. The size, location and possibilities of the towns are different. The same are planning traditions and attitudes. The solutions should suit local conditions. Therefore, the Intergovernmental Panel on Climate Change (IPCC) and most researchers consider it important that research methods continue to develop, which can safely measure these changes, improve the climate models and disseminate the information available to decision makers, so that they can make decisions regarding climate-adapted and risk-reducing measures [8]. Gardens, parks, green boreholes, forest and nature give the city quality. People are happier, less ill and less stressed when they can watch and move in the green. The green is the city's lungs, creates fresh air and provides shade and space for animal and plant life. Ample green city areas mean lower temperatures in the summer in the cities. The green provides opportunities for relaxation, to keep track of the season's shift and for play and movement [9]. We must think blue water is important quality in the city. Lakes, streams, canals, fjords and seas provide experiences and activities. The city will help ensure clean and abundant groundwater. Climate change means more water from above. It provides opportunities for more planned wetlands in the city. Our way of building a city is developing in a way where the energy consumption of cities is far greater than in the closer cities. We must return to the closer cities, thus creating less distance between the city's functions. Bike and public transport will be better alternatives to the car. It also offers better opportunities for collective solutions in other areas. It creates greater intensity and peace of mind, more life and more space for diversity. Therefore, we should think in greater density, no matter how large or small a city or district is. Higher density should not have negative consequences. Light, air and health are a natural part of the modern, sustainable city. The same are effective solutions for energy, water and waste. The "technology" must be alright—also in the city's buildings, where it is important to avoid dangerous substances affecting the environment and a health risk to humans. Climate change must be utilized positively. We should not only think close but also green and healthy. The EU countries' carbon dioxide emissions have fallen in all sectors in the first half of 1990s, except for one: the transport sector. But it is not from road traffic, as carbon dioxide has increased the most, but from air traffic, although flying still account for a lion share of total emissions from traffic. In total, EU countries' carbon dioxide emissions fell by almost 3% during this period. The decrease is partly due to the recession and the decrease in energy consumption in Germany when the DDR joined the Federal Republic. In addition, the British began using more natural gas and less coal [10]. Lack of water, polluted rivers, poorly functioning drainage systems or none at all have been so bad in many of the world's growing metropolitan areas.

- The rivers of Buenos Aires are clean sewers.
- In Karachi, southern Pakistan, 30,000 people die annually of polluted water.
- In Basra, southern Iraq, more than 4 million residences used frequently polluted water. More than 100 thousand residences are infected from polluted water.
- Shanghai faces threatening water shortage and saltwater penetrates the Yangtze River.

It shows a recent report from the World Nature Fund [11].

A couple of 100 million city residents lack access to a nearby source of safe drinking water. Chronic water shortages affect many more including rural residents. Approximately 50 countries of the world have too little water or are on the verge of a water shortage. In many countries, water shortages are already acute, such as in North Africa, the Middle East and parts of Asia [12].

Several European countries are also approaching a situation where clean water is insufficient. This applies to countries such as Poland, Belgium, Great Britain, Germany, Denmark and

Spain. Freshwater resources in the Third World are believed to be a central issue in international cooperation in the twenty-first century. Conflicts around water can be difficult to handle. It may be relevant to export water from countries with abundant supplies [11, 12]. The traditional way of managing water and sewage in the big cities is to get water further and further away and expand the pipelines. Then, the wastewater is pumped far away from the city. Mexico City is an example. Due to the over-extraction of groundwater, the city has fallen and had problems with flooding [11]. However, it is not certain that our Western system to manage water and sanitation is the best. It is in its place outdated with spent infrastructure, expensive and not adapted to today's needs and climate change.

3.3. An economically sustainable city

The cities are essential to the world economy. They are growth centers in a globalized economy, bringing economic development to the hinterland. The potential of the cities for attracting foreign tourists can also contribute to creating economic sustainability. The long-term development of cities must therefore also take into account tourism, which significantly contributes to growth and employment. Measures to ensure the economic stability of disadvantaged areas must also exploit endogenous economic forces within those areas. In this context, the labor market and economic policies specifically designed to meet the needs of disadvantaged areas will be the right instruments. The aim is to create and secure jobs and facilitate the creation of new businesses. In particular, opportunities to access the local labor market need to be improved by providing training that is tailored to market requirements. Also, in the context of the ethnic economy (generated by cultural diversity), employment and training opportunities need to be used to a greater extent. The European Union, Member States and cities are called upon to create better conditions and instruments for strengthening local economies and, at the same time, local labor markets, in particular by promoting social economy and by providing quality services to their citizens. It is useful in working with the concept of sustainability to understand and work with economics in a broader perspective that crosses social and green sustainability, as social and green sustainability costs play a role. By establishing a life-cycle perspective, you look at prices for both establishments, operation and settlements. For example, it is about how a better environment addresses the costs of cleaning or maintenance or how socially well-functioning urban areas give rise to fewer costs for repair, oversight and so on. If municipal politicians continuously focus on thinking public and private investments together, they can support each other and create added value in the investments. That way, high-quality service can be delivered in the most cost-effective way. In a long-term economic focus that embraces social and environmental factors, there are often much more significant gains to be anticipated for the benefit of both business and society in general. In other words, there may be growth and jobs in social and green sustainability as well as an export potential in sustainable urban solutions.

4. Recommendation

For a future arrangement of sustainable cities, the new urban agenda, adopted in Quito (Ecuador) on October 17, 2016, complements the 17 sustainable development goals assumed by world leaders at the end of 2015 through Agenda 2030 for Sustainable Development. The Strategic Vision of Sustainable Urbanization, negotiated for more than 3 years and presented at the Conference on Housing and Sustainable Urban Development (Habitat III), provides for a comprehensive approach to urbanization for the next 20 years. Thus, the sustainable urban development agenda proposed by UNDP is focused on concrete actions and sets global standards for sustainable urban development. It also includes a series of recommendations centered on rethinking how people will build, manage and live in cities by 2036 [5]. The economic and environmental footprint of urban centers is very high, despite the fact that they only cover a small part of the globe (0.51% of total land area globally). In terms of percentages, cities accounted for over 80% of world GDP in 2014, producing more than 70% of global greenhouse gas emissions and 80% overall energy consumption. At the same time, experts estimate that in 20–30 years the development of urban centers will be significant, tripling in size and reaching about 1.2 million square kilometers in 2030. Consequently, authorities are encouraged to use renewable energy sources (by 2040, the global energy system should meet the demand of 9 billion people), improve public transport, eliminate pollution and sustainably manage natural resources. According to the 11th sustainable development objective, focusing on the sustainability of cities and urban communities, two-thirds of mankind (around 6.5 billion people) will live in urban areas by 2050. That is why the way we build and manage our urban spaces becomes essential for our sustainable development policy. If 26 years ago mankind numbered 10 megalopolis, with more than 10 million inhabitants, in 2014 the number of these urban centers reached 28, totaling 453 million inhabitants. Thus, UNDP's goal in the next 20 years is to create safe and sustainable urban centers, which also aim to tackle challenges such as eradicating poverty, social inequalities and reducing climate change. The new urban agenda also addresses current issues such as urban poverty, unemployment, climate change, pollution, exclusion of marginalization of vulnerable groups (migrants and refugees), reduction of natural disasters and gender disparities. Even if the new agenda for urban development remains just a recommendation, UNDP recognizes that it needs to support national governments and local communities to meet the proposed objectives, the only ones that would favor the creation of more resilient and secure urban centers for citizens [5].

4.1. Back to the human scale interpretations

Quite large streets intersect many cities. It was considered 30–40 years ago to be the best answer to the challenge of rising car traffic. The roads lead, among other things, to large parking spaces in the city centers, which were built to ensure that people still wanted to shop on the city's main street when compared to pedestrian streets. Therefore, higher density is a significant challenge for many cities. Where city centers in the big cities are often characterized by high frequency, high versatility and a complicated life, many medium-sized cities have hollowed urban corridors that are quite desolate and insecure for much of the time. Paradoxically, the center in many cities is often the characteristics that characterize the periphery of the big city. These problems can be solved in many places by building closer. In some places, underground parking facilities or parking garages can be built. This gives new opportunities to looped parking spaces. Here, it is essential to mix housing, non-polluting professions, institutions, new activities and recreational green areas. This also creates a "green density" and stimulates health. Elsewhere, it may be better to think of public transport, so the need to park in the city centers is reduced. In this connection, attractive walking and cycling paths should be established, which connect the city center with its inner periphery. If people can go and cycle between houses, work, institutions and leisure facilities on beautiful and safe trails, more people will want to choose an excellent transport to the short distances. The edges of the city can be of great importance to health. One would think it was the most natural thing in the world to run or cycle the trip in green surroundings, at least for those living in a smaller city. But much of the land surrounding the cities is used for agriculture or other business purposes. When planning new housing quarters, it is essential from the start to think of healthy and recreational path systems into plans for the benefit of the entire city.

4.2. Applying for sustainable model theories in planning and design city processes

This requires an active application of the "Integrated Approach" idea into urban development processes, where integrated urban development policy is a simultaneous and fair consideration of all issues and concerns relevant to urban development. Integrated urban development policy is a process that coordinates critical spatial, sectoral and temporal matters. The involvement of economic factors, stakeholders and public opinion is essential. Integrated urban development policy is a vital status for implementing the UN's Sustainable Development Strategy. Its implementation of a requirement of a holistic world dimension, however, must take into account local conditions and needs based on the principles of subsidiarity.

Establishing a balance between the various interests of urban actors, supported by integrated urban development policy, is a viable basis for a consensus between the state, regions, cities, citizens and economic actors. Putting together knowledge and financial resources, public funds that are always insufficient can be used more efficiently.

Integrated Urban Development Policy involves actors outside the administration and enables citizens to play an active role in shaping their living environment. At the same time, these measures can provide more certainty regarding planning and investment. We recommend cities from all over the world to develop integrated urban development programs for cities as a whole. These implementation-oriented planning tools must have the following roles:

- describe the qualities and defects of cities and neighborhoods, based on an analysis of the existing situation;
- define realistic development objectives for the urban area and develop a coherent vision of the city;
- to coordinate technical and sectoral plans and policies related to the various regions of the city and to ensure that planned investments will help to promote a balanced development of the city and the surrounding area;
- to coordinate and to focus from space the use of funds by public and private sector;
- be organized at the local and regional level and involve citizens and other partners who can make a substantial contribution to the qualitative modeling of the economic, social, cultural and environmental future of each area.

Coordination between local and regional levels needs to be strengthened [13]. The aim is to establish at a regional and metropolitan level a balanced partnership between cities and rural areas as well as between small, medium and large cities. Urban policy issues and decisions can no longer be viewed in isolation at the level of each city. Our cities must be focal points for the development of the regions and take responsibility for territorial cohesion. That is why it would be useful if our cities were to work more closely in the network globally. Integrated Urban Development Policy provides a set of tools that have already proven their worth in many cities, which it can offer an efficient and cooperative management structures. These are indispensable for increasing the competitiveness of cities from the devolved countries. They facilitate early and timely coordination between economic, infrastructure, real estate and service development, taking into account, among other things, the impact of existing social trends on seasonal and population aging trends and energy policy conditions. To achieve the planned objectives and policy, it is important to consider that the following action strategies to be of crucial importance for improving the competitiveness of devoted world cities.

4.2.1. Greater sustainability requires both cultural and technical conversion

It is necessary to continue to safeguard the conservative buildings and cultural environments in the cities and promote new beautiful architecture. We must also create urban spaces that are strange and exciting to live in. Here, climate change can help us. More rainfall and warmer rainfall mean that more massive amounts of rainwater will be wasted when the large, fortified areas characteristic of the towns prevent rainwater from slipping into the soil [13]. Rain can be collected in artificial lakes or ponds that can prevent flooding. The reduction can be increased with more green areas, plantings and coatings that allow rainwater to sip down to groundwater. In this way, we can mitigate the effects of climate change while strengthening both the blue and the green flair of the cities. The city's buildings can also be used in the fight against climate change. It is technically possible to build houses, with sound building materials and houses that do not use energy but produce it. This is primarily about spreading knowledge about the type of construction.

4.2.2. Integrity and innovation in planning

Applying positive criteria of sustainability means creating sustainable cities is about reconciling and balancing many interests. The municipal reform has meant larger municipalities with many different urban communities. It can contribute to the fact that the cities increasingly have different roles, can complement each other instead of neighboring two former municipalities to compete and should be able to do the same. It requires holistic orientation, innovation and attitude changes.

4.3. Creating and securing respectable quality of public spaces

It is all the places in the community where there is access and where people are allowed to move freely outside the four walls of the home. There may be streets and alleys, educational institutions or meetings that may be more or less public. However, it is also the rooms that are carried by media of all kinds, which contain public debate and other expressions and where there is public access. Here, people may more or less be involved, and so it can be controversial. A prominent area is the public institutions where schools, kindergartens and jobs are found. The quality of public spaces, urban anthropic landscapes, architecture and urban development play an important role in the living conditions of city citizens. These local characteristics are important for attracting businesses in the knowledge industry, a creative and skilled workforce and for tourism.

4.4. Applying for green and healthy urban transformations

Trees and green areas have many advantages; they relax the stressed eyes. Trees and plants not only purify the air but also lower temperatures in the big cities in the summer months. Plants help to drain rainwater, thus reducing the need for other and perhaps more expensive solutions. Therefore, it does not play just a role in the urban transformation that should have a green dimension. Also, the existing city can be developed to make it greener and greener—for example, thinking of green roofs and green facades. Contrarily, changes in business structures create a natural need to transform past, often centrally located industrial areas into new types of jobs, for private and public services or for housing [14]. At the same time, there is also the need for a useful framework for the companies that remain in the city. When the major cities are transformed, it is essential to think about both the green and the health and to adapt the solutions to modern people. When laying parks and green fields on former industrial areas, it is essential that they are designed in such a way that the city's users and residents can quickly get a run or organize a ball game.

4.4.1. Objectives of using recycling materials model

Growing mountains of waste and eutrophication of the garden are two of the environmental problems that life in metropolitan areas can cause if politicians and civil servants do not work deliberately to end the cycle in the city. Working to make all residents easily sort their waste and leave it at places near the home is an important part of the material cycle. Well-functioning water treatment plants that take care of and clean the water from different chemicals and nutrients before it flows into the sea are also important. The sludge from the purification plants can partly be used in the production of biogas, and if it is sufficiently clean from poison, it is possible to return to farmland as manure.

These tasks require special attention in cities:

- Water and drainage
- Transport and communications
- Energy

4.5. Modernizing infrastructure networks and increasing energy efficiency

An essential contribution to improving living conditions, environmental quality and the creation of favorable factors for commercial locations can be ensured through sustainable, accessible and affordable urban transport with coordinated links to urban and regional transport networks [14]. Particular attention should be paid to traffic management and interconnection of modes of transport, including cyclists and pedestrians. Urban transport needs to be adapted to different housing, work, environment and public space requirements. Technical infrastructure, especially water supply, sewerage network and other urban networks, needs to be improved and adapted to changing needs in order to meet future demands and ensure quality in urban living. The key requirements for the sustainability of public utilities are energy efficiency, rational use of natural resources and economic efficiency in operation.

Analysis of possibilities for 2020 indicates that the developed countries can reduce their energy use by about 50%, while the developing countries with largely unchanged energy consumption per capita could reach a tangible standard equivalent to the one we had in Western Europe in the 1970s. The condition is that energy efficient technologies must be used in both developed and developing countries [15].

4.5.1. Energy efficiency in combater of climate change

The energy efficiency of buildings needs to be improved. This concerns both existing and new buildings. Renovating existing dwellings can have a significant impact on energy efficiency and on the quality of life of residents. Increased attention should be paid to prefabricated buildings, old buildings or lower quality buildings. Optimized and efficient infrastructure networks and energy-efficient buildings will reduce costs for both businesses and citizens. A basis for the efficient and sustainable use of resources is a compact structure of human settlements. This can be achieved through territorial and urban planning that prevents uncontrolled urban expansion through strong land supply and speculative development. Urban planning strategy to achieve a functional mix between housing, jobs, education and recreational use of urban areas has proven to be sustainable. Cities need to help ensure and improve the quality of life of their inhabitants and business attractiveness by providing sophisticated information and communication technologies in the fields of education, employment, social services, health and safety. Technical infrastructure, especially water supply, sewerage network and other urban networks, needs to be improved and adapted to changing needs in order to meet future demands and ensure quality urban living. The key requirements for the sustainability of public utilities are energy efficiency, rational use of natural resources and economic efficiency in operation.

4.5.2. Efficient use of renewable energy systems

The utilization of bioenergy, water and wind power can increase significantly. In the case of biofuels, it may be difficult to make the land sufficient for increased energy supplies, while the production of food and industrial raw materials must also increase and reach for about 10 billion people. Solar cells in desert areas could be an opportunity to produce electricity. Major and urgent efforts are needed to develop the technology needed for renewable energy laws. Various global scenarios for energy supply in 2050 have been made by, among others, the World Energy Council (WEC), the World Watch Institute and the Intergovernmental Panel on Climate Change (IPCC). In all scenarios, fossil and renewable fuels are used. Most of the scenarios still have nuclear power in the energy balance of 50 years, but the share of nuclear power of total energy supply is less than today. Much of fossil fuels will still be used. The natural gas dominates coal and oil in all future images and, according to the World

Watch Institute, is by far the only fossil fuel remaining in 2050. The energy scenarios do not determine whether they are compatible with sustainable development or not [16]. In general, the vision of a multifaceted architecture shows that it is necessary to design and work cooperatively with an architectural theory to transform the lateral conceptual viewer of the multilateral design process [17].

4.5.3. *Efficient use of local energy systems*

Local energy systems can be used in many places. Energy system means the recovery and storage of energy in the form of heat from the ground. The design of the building may be controlled by any local energy system. For faced the future energy shortage, it is required to recognize all survived opportunities. In this context, it is necessary to consider all existent possibilities [15, 16].

4.5.4. *Efficient uses of geothermal energy*

Geothermal energy is the heat energy stored in the Earth's crust and flows out to the ground. The heat comes mainly from radioactive decomposition. The geothermal gradient varies from place to place. In Nordic countries, such as Sweden or Denmark, the sedimentary bedrock has the best conditions, and heat can be recovered using deep drilled wells [15, 16].

Soil heat utilizes heat energy from the sun and rain stored passively in the ground. The heat is absorbed through plastic hoses laid down at 1–2 m depth in which a freeze-dried liquid circulates. The heat is then recycled using a heat pump. To heat a small house, a surface area of 400–600 m^2 is required. If the hoses are laid above the groundwater surface, fine-grained soils that hold the moisture are well suited. Silt should be avoided due to the risk of fire. If the hoses are laid under the groundwater, water is important, so that coarse grains are preferable [15].

4.6. Proactive innovation and educational policies

Cities are places where knowledge is created and shared. Exploiting the full potential of knowledge in a city depends on the quality of day care and school education, the transfer of opportunities offered by education and training systems, social and cultural networks, opportunities for lifelong training, the excellence of university education and research institutes and the existing transfer network between industry, the business sector and the scientific community. Integrated urban development policy can help improve these factors, for example, bringing together all stakeholders, supporting networks and optimizing infrastructure. Integrated urban development promotes social and intercultural dialog [18]. Integrated urban development strategies, participatory urban management and good governance can help to make effective use of the potential of European cities, particularly in terms of competitiveness and growth, and to reduce disparities between and within neighborhoods. These integrated urban development strategies bring citizens the opportunity of social and democratic participation.

4.7. Applying for new business for new opportunities

Business development in the last two to three decades has been of major importance for urban development. Retail trade is an important factor in urban life. Wherever shops are placed, they dramatically affect traffic and trade patterns. Large sections of the industry have moved or disappeared, more may follow in the coming years. The abandoned industrial areas at the ports, in the city center or in their periphery open up opportunities for building closer and greener for the benefit of both city life and sustainability. Shipping and fishing have also changed; therefore, in many port cities, it is necessary to consider how the areas can be recycled. Even though many jobs have been closed, even population of people has increased. It is typical in industries that do not pollute or disrupt. There are industries that can be more easily mixed with, for example, housing, and can contribute to a good urban environment. This makes it necessary to consider whether existing rules and planning traditions to regulate coexistence between businesses and housing are still appropriate [19]. There are also professions that necessarily have to be in the cities and in the ports. Many of these companies are venture companies that cannot be instantly integrated with homes and other common urban activities. It is essential to ensure that these companies can continue to function and develop, and it must be ensured that there are sufficient safety distances around these companies—for the sake of both the companies themselves and their neighbors. It also means that the ports should not only be emptied for business and converted into, for example residential areas [20]. There is still a need for ports and shipping.

4.8. Paying attention to underground construction concept

In many countries such as Japan, Malaysia and China, planning to take advantage of surfaces of the inner parts more effectively have begun in these countries. A solution for creating a sustainable future for the Earth's growing population is building on several floors, that is, to use surfaces above and below ground level. The compact city is seen as a model for future low-energy society. Japan, for instance, are looking at building whole cities down the mountain. The most important subject which required a clear answer is a way to cope with the transport problems in the largest cities. In order to accommodate all the facilities needed in the metropolitan areas, underground construction is being increasingly used for warehouses, business centers, shelters and facilities for work, leisure and cultural activities [21]. There are obvious environmental benefits. When an underground project is to be evaluated in an environmental impact assessment way, the aforementioned option should also be evaluated environmentally.

4.9. Paying particular attention to "negative areas" in the context of the city as a whole

Cities face major challenges, particularly in connection with changes in economic and social structures and globalization. Specific issues, among others, are a high unemployment rate and social exclusion phenomena. Within a city, there may be considerable differences not only in terms of economic and social opportunities but also in terms of the quality of the environment between the different areas of the city. In addition, social inequalities and differences in levels of economic development of the areas continue to increase, which contributes to the imbalance. A policy of social integration that contributes to reducing inequalities and preventing social exclusion will be the best guarantee for maintaining security in our cities. In order to achieve the objectives of social cohesion and integration within cities and urban areas, a good conception of housing policies is needed [22]. It is better to heed early warning signals and to take immediate and effective measures to remedy the situation as this approach saves resources. If an area has begun decaying, the cost and difficulties of returning that area to the waterline may often be higher. The government must provide the general framework and rehabilitation incentives for the inhabitants of the affected areas. Active involvement from residents and a better dialog between politicians, residents and economic actors are essential to finding the best solution for each disadvantaged urban area. It is believed that the following action strategies, included in an integrated urban development policy, are crucial for deprived urban areas.

4.10. Remarkable attention on sustainable solution of "Transport in the city"

The world's largest cities have very bad air quality. Traffic jams are common in cities worldwide, including the majority of the in some countries. In big cities, it is common that people travel several hours daily in order to access the workplace. The level of motoring has in many places reached a point where the problems are no longer possible to accept. Today, there are hundreds of millions of motor vehicles in the world, and many hundred thousands of people are killed on the roads each year [23]. In some countries, the goal is for every household to get a car. The trend in many countries is that the number of cars increases as income increases, households are divided and cities are expanding. Another problem is the poor connectivity between human settlements due to the poor quality of road infrastructure. Agenda 2030 addresses these issues by upgrading the transport system so that it can better connect cities and settlements and make it affordable, including financially and durable, to ensure that all world citizens benefit from the urbanization process. The focus is also on increasing access to green and public spaces, reducing the negative impact of cities on the environment per capita and reducing direct economic losses caused by disasters.

4.10.1. Suburbans have to connect with the cities objectively

The oldest suburbs were based on modes such as trams, city busses and trains. The newer suburbs are often based on the car as the dominant mode of transport. The vehicle is a minimal sustainable mode of transportation—especially if it only has the driver on board. The most of the crowds on the landing roads and in the big cities come from the suburbs. The suburb of the future must also be based on modes other than the car. Suburbs can be connected to the city center with light railways or modern trams that can transport a lot of passengers with a modest width. In other places, frequent and fast buses can reduce the need for driving. Attractive bicycle parking at stations and stops can encourage more to use bus or train.

4.10.2. Local stations in the suburbs

However, the vehicle cannot be avoided. On the other hand, it can be used more appropriately. Municipalities with suburbs can build larger parking spaces at the local stations in the suburbs, thus encouraging the drive not to go all the way to work or to the city center. They can promote partial arrangements and intercourse, resulting in fewer cars on the streets.

4.10.3. Using of subways as environmental solutions

The interest in concealing road traffic is increasing in the century with environmental requirements. One way is to build tunnels. Many of the world's most complex tunnel projects are conducted in Asia's fast-growing metropolitan areas. There are many examples of traffic tunnels in several floors, subways, tunnels for district heating, telephone lines and so on. Subways are expected to be very important in the future. In many areas, there are also plans for car traffic in tunnels, including in Stockholm. All technical systems that utilize pipes or pipelines are built underground for its distribution network: telecommunications, electricity, district heating, gas, water and sewage [24].

4.10.4. Promote efficient and cheap urban transport

Many disadvantaged neighborhoods also lead to the lack of transport links and environmental influences, which are likely to reduce their attractiveness. The development of an efficient and inexpensive transport system will provide residents of these neighborhoods with opportunities for mobility and equal access to those of other citizens. To achieve this, transport planning and traffic management in these areas must progressively reduce the negative impact of transport on the environment and organize transport in a manner that integrates these neighborhoods into the city and region as a whole. Suitable networks for pedestrian traffic and cyclists will be useful for this purpose. The better we manage to economically stabilize disadvantaged areas, integrate them socially and improve the physical environment and transport infrastructure, the greater our chances are that our cities will remain points of social progress, growth and innovation [25].

4.11. Activation of training education policies for children's and young's residences

A crucial starting point for improving the situation of disadvantaged areas in cities is to improve the situation of education and training in local communities in line with proactive policies focused on children and young people. More opportunities for education and training should be provided for disadvantaged areas, and they must be tailored to the needs and deficiencies of children and young people in those areas. Through a policy focused on children and young people built to the requirements of the social area in which they live, we must contribute to increasing the chances of children and young people living in deprived areas to achieve and provide equal opportunities in the long run.

Author details

Amjad Almusaed[1]* and Asaad Almssad[2]

*Address all correspondence to: a.amjad@archcrea-institute.org

1 Department of Construction Engineering and Lighting Science, Jönköping University, Sweden

2 The Head Department of Building Technology, Karlstad University, Sweden

References

[1] Cucu V. Geografia orasului. In: Fundatiei "Dimitrie Bolintineanu". Romania: Bucuresti; 2000. p. 30

[2] Almusaed A. Introductory chapter: A general reading process on landscape architecture. London, England: Intechopen; p. 3-7. [Accessed: September 19 2018]. DOI: 10.5772/intechopen.77971

[3] Slavin MI, editor. Sustainability in America's Cities Creating the Green Metropolis. Washington, D.C., United States: Island Press; 2011. p. 2

[4] Almusaed A. Intelligent Sustainable Strategies Upon Passive Bioclimatic Houses: From Basra (Iraq) to Skanderbeg (Denmark). Aarhus, Denmark: Aarhus School of Architecture; 2004. p. 154

[5] United Nations. New Urban Agenda, Habitat III., Supported by the Government of the Republic of Ecuador. 2017. p. 3, 4, 7, 41

[6] Almusaed A, Almssad A, editors. Sustainable Building, Interaction between a Holistic Conceptual Act and Material Properties. London, England: IntechOpen; 2018. p. 5

[7] Baker S. Sustainable Development. New York: Routledge; 2006. p. 26

[8] Bojinski S, Doherty SJ. Lessons learned from IPCC AR4 [internet], American Meteorological Society. 2009. Available from: http://journals.ametsoc.org/doi/abs/10.1175/2008BAMS2643.1

[9] Almusaed A, Almssad A. Urban biophilic theories upon reconstructions process for Basrah City in Iraq. In: The 30th International Plea Conference, December 16-18, 2014. Ahmedabad: CEPT University; 2014

[10] Almusaed A. Towards a zero energy house strategy fitting for south Iraq climate. In: The 25th Passive, and Low Energy Architecture (PLEA). October 2008. Dublin, England: University College Dublin; 2008

[11] Engel K, Jokiel D, Kraljevic A, Geiger M, Smith K. Big Cities. Big Water. Big Challenges. Water in an Urbanizing World. Berlin: WWF Germany; 2011

[12] Zoomers A, van Noorloos F, Otsuki K, Steel G, van Westen G. The rush for land in an urbanizing world: From land grabbing toward developing safe, resilient, and sustainable cities and landscapes. World Development. 2017;**92**:240-2525

[13] Almssad A, Almusaed A. Environmental reply to vernacular habitat conformation from a vast areas of Scandinavia. Renewable and Sustainable Energy Reviews. 2015;**48**:825-834

[14] Almusaed A, Almssad A. Building materials in eco-energy houses from Iraq and Iran. Case Studies in Construction Materials. 2015;**2**:42-54

[15] Harris AM. Clean Energy: Resources, Production and Developments. New York: Nova Science Publishers, Inc. (Energy Science, Engineering and Technology); 2011

[16] McNerney G, Cheek M. Clean Energy Nation: Freeing America From the Tyranny of Fossil Fuels. New York: AMACOM; 2012

[17] Almusaed A, Almssad A, Alasadi A. Analytical interpretation of energy efficiency concepts in the housing design process from hot climate. Journal of Building Engineering. 2019;**21**:254-266. DOI: 10.1016/j.jobe.2018.10.026

[18] Almusaed A, Almssad A. Effective Thermal Insulation, The Operative process of an efficient Passive Building Model. Croatia: Intech Publisher; 2013. p. 20

[19] Kellner J. Housing Reclaimed, Sustainable Homes for Next to Nothing. Gabriola Island, BC, Canada: New Society Publishers; 2011. p. XI

[20] Almusaed A, Almssad A. Biophilic architecture, the concept of healthily sustainable architecture. In: The 23th Conference on Passive and Low Energy Architecture. September 2006. Geneva, Switzerland, PLEA: Geneva University; 2006

[21] Shiftan Y, Attard M. Sustainable Urban Transport, Transport and Sustainability. Bingley, UK: Emerald Group Publishing Limited; 2015

[22] Jacobs D et al. A systematic review of housing interventions and health: Introduction, methods, and summary findings. Journal of Public Health Management Practice. 2010;**16**(5):5-10

[23] Gehl J. Cities for People. Washington, DC: Island Press; 2010

[24] Almusaed A. Biophilic and Bioclimatic Architecture, Analytical Therapy for the Next Generation of Passive Sustainable Architecture. London, England: Springer-Verlag London Limited; 2011. p. 198

[25] Campbell Lendrum D, Corvalan C. Climate change and developing-country cities: Implications for environmental health and equity. Journal of Urban Health: Bulletin of the New York Academy of Medicine. 2007;**84**(1):109-117 (Jordan pilot project)

Section 2

Energy and Environmental Analysis of Sustainable Cities Models

Chapter 2

Towards Adaptive Design Strategies for Zero-Carbon Eco-Cities in Egypt

Marwa Dabaieh, Dalya Maguid and
Deena El Mahdy

Additional information is available at the end of the chapter

http://dx.doi.org/10.5772/intechopen.80725

Abstract

Eco-cities concepts are relatively new initiative launched by the World Bank, to help cities in developing countries realise ecological, social and economic sustainable future. Furthermore, with growing severe climatic events such as the rise in global temperature, flooding, wild land fires, and sea level rise, there is an urgent need to adopt sustainable and ecological design principles for the development of future cities. Egypt, one of the developing countries and third largest populated nation in Africa, is currently facing a series of threats. These include limited access to natural resources in relation to the population size and economic growth. In addition to the continuous challenging climate change implications. Despite that, till now there are no clear laws or legislation for eco-city design and construction. In this manuscript, we are trying to lay hands on hidden potentials and analysing successful private initiatives for existing eco-communities in Egypt. We adopted the analytical case study method tackling different aspects like renewable energy, permaculture, eco-sanitation, solid waste management, vernacular architecture, green transportation and green economy. The research contributes by critically analysing such attempts and concludes with design recommendations and strategies on how to reach an environmentally enriched, healthier, resilient and socially rewarding zero-carbon cities, running on their own locally available resources for the Egyptian cities.

Keywords: zero-carbon eco-cities, urban future, green urbanism

1. Introduction

Cities alone account for 78% of anthropogenic carbon emissions [1]. In the developing world, cities produce and consume at higher rates than rural areas per capita [2], and accordingly account for an unbalanced share of greenhouse gas emissions. Let alone that the majority of urban south cities with coastal location with high concentration of economic activity and population [3] are the most vulnerable to the effects of climate change. In short, cities and consequences of human settlements are at the core of the problem, both in terms of carbon emissions at sources, and in terms of the effect of global warming. Low- and zero-carbon city design and planning can play a crucial role in reaching global targets in CO2 reduction, ensuring long-term energy security and reducing the impacts of climate change worldwide [3].

Egypt like many other several developing countries is facing a series of threats related to limited access to natural resources in relation to the population size and economic growth and energy poverty. All that had a strong influence on urban development and planning. Directing future development out of the narrow Nile valley and Delta becomes indisputable. The future urban development plans in Egypt for 2050 aim at spreading development over 40% of Egypt's area to exploit available natural resources and provide around 20 million job opportunities [4]. Other key concerns now in Egypt are climate change and moving towards a zero carbon era. This is a growing concern with Cairo being the most polluted capital in the Middle East, and according to the World Health Organisation [5], the second most polluted large city in the world in 2018. However, the majority of researchers and designers now throw around the terms eco-cities and eco-economy without clues on how to achieve them.

On a very limited scale in Egypt, few pilot settlements were early focusing on environmental sustainability, using renewables and eco-friendly building construction methods which have given such settlements a vital edge within national competitive sustainable communities. The majority however are private initiatives like Basata in Sinai, Tunis in Fayoum, El Basaysa in Sharkeya and New Basaysa in Ras Sedr. Only one so far is a governmental initiative named productive, low-cost and environmentally friendly village (PLEV) which is still in its study phase. When it comes to rating systems and assessment tools to achieve the objectives of the eco-cities, unfortunately the Egyptian rating system the Green Pyramid is still on hold, and even if it reached the action stage, it did not include any chapters for sustainable communities or eco-cities.

This study comes aligned with the new government decision for building a new capital city in new Cairo and another millennium city called New El Alamein. The proposed designs are targeting the wealthy minority and prioritise self-determination without looking at affordability, low-impact living and equality. Investing in green infrastructure especially when it comes to water conservation together with applying energy-efficient building design strategies was not of a concern. Hence, there is a great demand and need for a discrete code or guideline to help in designing and assessing new or existing cities. There is also a necessity for transdisciplinary eco-guidelines for our contemporary and future eco-city design and planning. But how can we design a community that is both liveable and sustainable is a key question that this study is trying to answer by filling in this gap in national codes and rating systems and showing broad lines for a pathway towards zero-carbon eco-city design in Egypt. It draws upon three case studies for private eco-communities in Egypt which are considered as pilot initiatives. Here we are analysing the three case studies and showing different design elements and drawing recommendation that can be used as a set of design framework for an eco-city model in Egypt like renewable energy, permaculture, eco-sanitation, solid waste management, vernacular architecture, traffic and green transportation, green economy and small/micro enterprises. Basata,

Habiba and El Gouna eco-communities were used as examples chosen as pilot projects, and they represent serious attempts in applying low-impact strategies and zero-carbon principles in sustainable building and community design. The three cases were also selected because they represent unique examples for an eco-community with an underlying philosophy of community behaviour and cultural understanding. They are designed, developed and managed in an environmentally sensitive manner. The main common philosophy was devised in response to the desire to preserve the nature with minimal carbon footprint. This analytical study contributes by drawing a vision of what a future sustainable zero-carbon eco-city in Egypt could look like.

2. Case study methodology

The study applied qualitative comparative analysis approach using case study methodology. The methodology comprises an in-depth literature search for previous work on eco- and zero-carbon cities to scan the field and understand where the Egyptian situation lays within the modern approaches in eco-cities notions. In this phase, we took a look at current policies of city development in Egypt, especially for the new capital and new El Alamein city. We also scanned for existing design and planning notions and looked at current examples to study. Three cases were then selected and analytically compared to each other to deduce lessons learned for possible recommendations for future city designs.

The three case studies represent private initiatives for pilot models for eco-communities in Sinai Peninsula in the northeast part of Egypt. They were chosen because they show a good mix of low-tech and high-tech building methods using available local materials. They also represent community-driven versus business-driven design approaches. During the case study investigation process, we carried out several site visits and interviews with the projects' owners and some of the workers and inhabitants/visitors. That was one main tool to collect the needed information and materials needed for the study. The projects managed to apply an array of low-impact and sustainable core principles for eco-communities but not necessarily that they have succeeded in all. As they are located in the same geographical zone, they also share the same climatic zone which is mild costal arid climate with harsh summer season. It was easy to compare the climate responsive solutions if any. We have picked the successful best practices and analysed them to drive key design applications that can be considered a start for a pathway towards eco-communities in Egypt. The aim is reach to a set of recommendation for principles that are low-cost, low-impact and can be community-driven and managed.

3. The four pillars of sustainability

As a known aim, sustainable development looks to improve the quality of life while preserving the Earth's natural resources [6]. There are three main pillars for sustainable development, which were first defined during the Development Congress in Johannesburg in 2002 and were later on further enhanced by scholars. These are the social, economic and environmental pillars [7–9]. Economic sustainability calls for integrated approach that allows long-term growth while ensuring that no nation is left behind [6]. The utilisation of resources should not affect future income, allowing equity of resources for all generations, distributional equity and economic activity that are concerned with ecological aspects [9–11]. In terms of environmental sustainability, this refers to preserving natural resources for future generations and ensuring that natural resources are well managed and are being used with a suitable rate that would allow for regeneration of resources for the future [9, 11]. Social sustainability involves social cohesion [12], continuity of social values, identities and relationships and the sustainable presence of health, education, food, water, healthcare, and housing for all people [6, 9].

Moreover, recent literature indicates the importance of adding a fourth pillar for sustainable development, which is culture [13]. Cultural sustainability has seven main aspects that should be considered: heritage, vitality, economic viability, diversity, locality, eco-cultural resilience and eco-cultural civilisation [13]. Culture has previously been considered a part of social sustainability, including aspects such as equity, participation, awareness of sustainability, behaviour and preservation of sociocultural patterns, social capital, social infrastructure, social justice and equity [13]. However, culture is not yet included; thus, some scholars consider social and cultural aspects to be closely interconnected, as cultural values can influence social life. Furthermore, culture can actually be viewed as a necessary condition upon which all other aspects of social, economic and environmental sustainability can take place [13] (**Figure 1**).

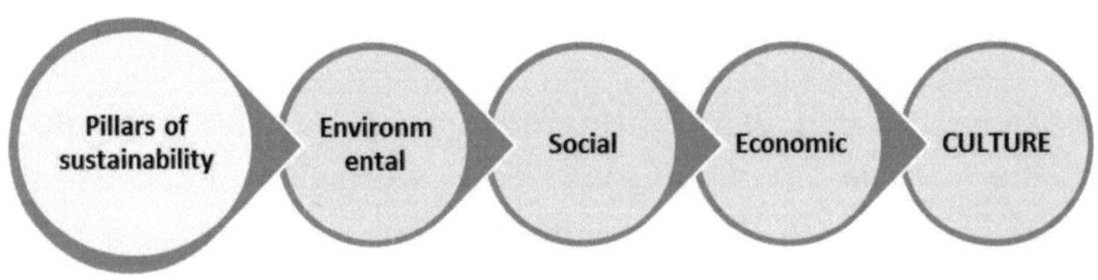

Figure 1. The four pillars of sustainability.

4. Eco, zero-carbon and productive cities: history and development

Various academics have put together definitions and theories for an eco-city. The term 'eco-city' was first coined by Richard Register to describe a city where human beings can exist in harmony with nature therefore greatly reducing our ecological footprint [14]. It is also described as a combination of many innovative design and planning ideas, each of which complement each other to form a reliable, amalgamated public environmental and climate-responsive community [15]. In an eco-city, science and industry collaboratively work hand in hand to achieve technical innovations, quality development and strong, long-term employment of its inhabitants [16]. It is a comprehensive and transdisciplinary concept and an amalgamation between science and technology, municipal policy and responsible citizenship [15]. It had been discussed in discourse that management of the environment is a key factor in local economy of eco-cities concepts together with other decisive factors such as culture, climate, and landscape coupled with locals' lifestyle and their ambition for a better quality of life [17]. An eco-city is much more than a few buildings that are energy-efficient. It is a complete design package that examines every element of human interaction with nature and positions that interaction in a way that humans have much less impact on the activities of other life forms [18].

http://dx.doi.org/10.5772/intechopen.80725

Over the last decade, both researchers and academic have been arguing around interdisciplinary concepts of eco and sustainable cities. There are many wicked and interconnected challenges like fuel poverty, climate change and ecosystem degradation intensively discussed in the discourse. Many have drawn concerns about low-impact design approaches and how to make low-carbon transition and reduce carbon footprint [19]. Others focus on eco-urbanism, while many others tackle the notions of zero-carbon and recently post-carbon together with finding ways to understand new forms of carbon on calculations through carbon value change [20]. Nevertheless, researchers had pinpointed that towards a road map for post-carbon cities, we should adopt multiple solutions than can work on different scales, starting from planning fundamentals till building a sense of community and calling for mainstream policies of urban development [21]; others highlighted the importance of community-led approaches towards low-carbon urban governance [22] and more chances for grassroots policies [23]. In the end, the idea is how to design a community that can become low- to no carbon starting from construction phases till operation. Furthermore, to be able to depend mainly on renewables in producing energy while taking into consideration other main key principles like transportation and land use and energy consumption. It is becoming evident now that societies with less dependence on fossil fuel resources decarbonise their energy, necessitate occupants' behaviour change and value perception. In addition, more effort is needed to adapt new technological innovation for energy saving [24]. That also should not be neglected towards defining a clear road map for zero-carbon societies.

The concept of productive cities introduces the idea that cities integrate several approaches that would allow them to generate their own energy and resources. This could include water, food and energy production. This is becoming quite important with the growing increase in food and water security problems all over the world, due to agricultural land being converted to residential and industrial areas [25]. Furthermore, after the industrial revolution, infrastructure systems increased, and many transportation systems improved, ultimately leading to greater urban growth. This allowed distances to be less restrictive, and cities no longer had to be situated next to their needed resources [26]. As a result, more cities have depended on importing their raw materials, food and energy, using them inefficiently and producing a great amount of waste. Moreover, there has been an increased dependence on fossil fuels for energy, which are being consumed in a rate that is much higher than their generation [26]. Thus, in order to ensure sustainability, it is imperative that future cities become places of production [25]. A sustainable urban planning approach is needed that would enable inhabitants to make use of available local renewable resources and be productive of their own food and energy [25, 26].

In terms of food production, there has been growing attention to the concepts of productive urban landscapes and urban agriculture [27, 28]. Productive urban landscapes are all open urban spaces that are planted in a way to be productive either environmentally or economically. This could include their use for food production from urban agriculture, removing pollutants from the atmosphere, maintaining biodiversity or simply improving the urban microclimate from trees [27]. Urban agriculture is mainly concerned with the growing of food and raising of animals for food in areas within or around the city [29]. Food growing can be on the ground, on roofs, in the building facades as vertical green walls or at fences and boundaries [27]. Recent research indicates that this can actually have several benefits for cities. This includes enhancing urban food security, providing job opportunities, urban greening of the city and giving an opportunity to use urban organic waste as a resource [29]. Compost production, vermiculture and irrigation are all examples of using urban wastes into productive resources [29]. Mougeot [30] also asserts that it is a vital part of the urban economic, social and ecological system.

Furthermore, productive urban landscapes can be in the form of vertical landscapes as well [25, 27]. Vertical landscapes include the placing of vegetation vertically against the building façade. This can act as a second skin that may not only be used for food production but also can have environmental benefits as well [27]. This will be made easier in high-rise towers through the use of photovoltaic cells that would allow vertical farms to be self-sufficient and sustainable. This is because providing sufficient lighting to mimic the sunlight and water pumping for irrigation are usually the main energy concerns in vertical farming [25, 31]. In terms of energy, Leduc and Van Kann [26] proposed a sustainable urban energy planning approach to create productive urban regions. Leduc and Van Kann [26] concluded that urban harvesting should be used as a planning approach, to transform urban regions into more productive areas. This includes taking into consideration local materials and resources and harvesting them to minimise any imports, while waste is identified as a valuable asset. Furthermore, resource consumption patterns in cities are closely linked to urban functions. Therefore, description of urban functions is indicated as an important aspect that can help provide valuable information about local resource demand and availability [26].

5. Three case studies: description and analysis

We start here by introducing Basata, Habiba and El Gouna, the three cases studies chosen for this analytical part. Basata is an eco-lodge for tourists' activities with a main goal of supporting the Bedouin local community located in Nuweiba in South Sinai, Egypt. The eco-lodge includes main residential units and an educational centre that gave the chance to the local Bedouins to have access to basic education which is lacking in that remote part of Sinai. It attempts to create a special self-sustained eco-community around the main activity of tourists' units (huts and chalets) that has a relatively low impact on the surrounding environment and the native inhabitants (**Figure 2**).

Habiba is an agricultural-based community farm located in Nuweiba as well. The community is composed of a beach eco-lodge, an organic community farm and a learning centre that

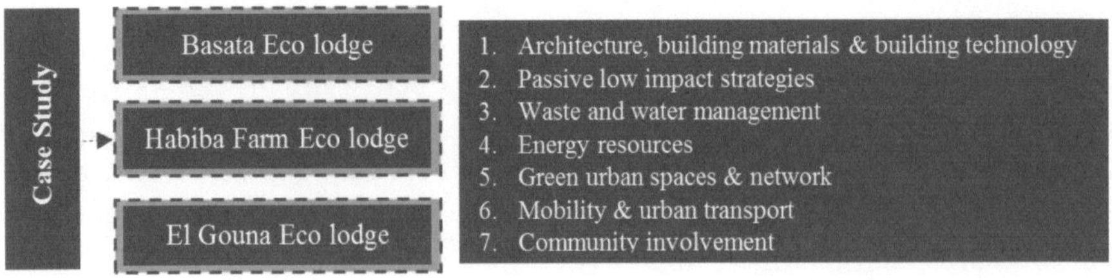

Figure 2. The three case studies analyzed in this study.

provides after-school education for local Bedouin children. The Habiba community advocates the idea of 'agritourism' and fosters the idea of international exchange in this aspect. It encourages having tourists and international specialists to come stay with them for a few months to share their experience in organic desert agriculture or to conduct research and experiments. Furthermore, the community also created the 'Sinai Palm Foundation', which is concerned with the expansion of palm date agriculture in the Nuweiba area. Moreover, the Habiba community partakes in different endeavours with partners, organisations and universities from all over the world to help in building their sustainable desert community in Nuweiba.

As for El Gouna, it is a small community spreads over 10 million m^2 of unspoiled terrain with 10 km of pristine beachfront located on the coast of the Red Sea, 25 km north of Hurghada. It contains residential area such as hotels, resorts and housing area together with public buildings such as universities, schools and libraries and finally the commercial and services area. On the other hand, El Gouna is to a great extent a gated community, with guarded entry gates surrounding the community. It is divided up to several parts; the island of El Kafr, the downtown area, the Marina area, and the workers' residential quarters. El Kafr mostly holds residential units. The downtown area is the main commercial zone, devised of several shops, restaurants, low profile residential areas. The Marina area holds two artificial bays with a waterfront walkway that has several small buildings (**Figures 3** and **4**).

Figure 3. The urban layout and the arrangements of huts and chalets in Basata showing the footprint and low density.

Figure 4. To the left sample of architecture style at EL Gouna and the right for the trials in using natural materials in Habiba.

5.1. Architecture, building materials and building technology

The main materials used in Basata's residential units are natural widespread biodegradable materials in Egypt like reeds, clay, straw and natural stones. The idea of using mainly using reeds, baby bamboo, straw and clay is to reduce the embodied carbon in the construction and also in lifetime energy usage. The applied indigenous construction methods reflected the local vernacular architecture of the region. The design was a response to cultural and social aspects as well. The main notion is emphasising the special character of the desert architecture and avoiding using any materials that can harm, pollute the environment or have high embodied energy together with eluding the use of any heavy construction equipment to avoid noise pollution during construction. For both El Gouna and Habiba, there are serious trials in using natural materials mainly stone, wood and clay, but the majority of the structures in both are manually using conventional materials like cement and fired bricks. In Habiba, they had examples of rammed earth construction, buildings with straw and buildings with recycled bottles. Many other building techniques were implemented as research work, as researchers can experiment and build their structures for testing and experimentation. In the three projects, stones and local materials are used in the hardscape like pathways and benches (**Figure 4**).

5.2. Passive low-impact strategies

In Basata, several passive strategies were applied for energy-efficient cooling, heating, ventilation and natural daylight. The optimal use of natural wind patterns and night flush effect together with encouraging cross ventilation reduced the need for air-conditioning or any other forms of artificial cooling in summer especially inside the reeds and bamboo huts. While in the adobe/stone chalets, the high thermal mass of the walls and high-domed ceilings played a major role in reducing heat gains during summer and providing adequate indoor thermal comfort all year round. In winter, for example, the warm night sea breeze and thermal mass of the chalets stone walls help in providing warm indoor comfort with minimal need to night heating. In Habiba, the applications for passive systems were limited, but there were few trials in using passive heating systems like Trombe walls, and for cooling they depend mainly on cross ventilation strategies using windows and doors crossing each other. In El Gouna, they mainly depend on high thermal mass walls. In addition to some roof structures that are dome or vault shaped with high ceilings, which help in reducing cooling demands during the long summer season. Wind catchers and solar chimneys were used in some of the buildings for cooling and ventilation. The same for using passive cross ventilation which is crucial in the three projects because of the hot humid weather (**Figure 5**).

Figure 5. Example for using wind catches as passive cooling strategy in EL Gouna.

5.3. Waste and water management

Waste sorting has a strict policy in the three projects. In Basata and Habiba, organic food excess from cooking and from meals left over are used to feed animals grazed on site. Animal manure is used as an organic natural fertiliser for the plants, and it is also used in the clay soil mixture in building. For both Basata and Habiba, all solid recyclable materials are sorted and picked by a local NGO to the solid waste transfer station in Nuweiba, where they are further sorted and sent for recycling in main factories in Cairo. While in El Gouna, all waste is also recycled and sent to main waste stations in the nearby area. In the three projects, there is a water desalination plant onsite that produces fresh water and brine water (highly concentrated salt water). Also, they apply strict water conservation policy. Fresh water is used in kitchens only for cooking and rinsing the dishes, while in the bathrooms only for showering and handwashing. Brine water is used for the rest of purposes like washing the dishes, toilet flush and for construction work. In Basata and on some parts in El Gouna, toilets water saving taps are used and that helps in reducing water consumption. On average, each person uses from 70 to 100 l of water per day, compared to an average of 500 l per person per day in neighbouring tourism communities. As for wastewater, it is divided into grey water and black water. Grey water is used for irrigating nonedible crops, endemic palm trees and plants. The salty black water goes into sealed septic tanks and then is transferred to the main water treatment plant. In Habiba, the eco-lodge promotes the ideas of water management and conservation. A waste recycling system is also implemented in the eco-lodge but no waste water treatment system.

5.4. Energy resources

Basata and Habiba are located in an off-grid site so as an essential need; they had to start with diesel generator as renewable energy which was not feasible in terms of high cost. Now in Basata, they started gradually to replace the generators with solar panels for electricity and solar heaters for water heating. Energy saving light bulbs are used. In some building units, there are basic electric equipment, but no air-conditioners are installed, neither are TV sets, refrigerators nor electric entertainment facilities. In winter time, they tend to turn off the electricity generators for several hours during the day, since kitchen refrigerators are not in use as much. One of the strategies to reduce light pollution on site is to minimise the outdoor lighting features. For the Habiba community, it is concerned with energy and resource preservation. In the eco-lodge, there are no electric water heaters for energy conservation, and roof top coolers are used. However, the place does not integrate the idea of producing energy whether through solar cells or any other renewable means. At El Gouna, some of the buildings use solar water heaters located on the roof tops. Also, they have photovoltaic cells for electricity production in some buildings.

In both Basata and Habiba, food production and urban agriculture are, however largely implemented through the community organic farm. This provides inhabitants and locals with the locally produced organic food thus helping in creating a self-sustained community in terms of food production. In Habiba, the focus is more on the agriculture compared to the other two projects (**Figure 6**).

5.5. Green urban spaces and networks

In Basata, specifically, the approach for site planning aims for satisfying living needs and changing radically to a more efficient use of land. The site planning reduces the ecological

Figure 6. Using solar water heater and PV systems for electricity production in EL Gouna.

footprint with a minimal level of urban density. It is easy to cycle or walk and move around the place even when there is no specific cycling or walking lanes. One of the main environmental site concerns was but is not limited to preserving the abundant marine life, fauna and flora. There is no artificial planting or importing non-endemic flora. Grass is not used as ground cover because it requires huge amounts of freshwater and polluting fertilisers.

While in Habiba farms, the green urban spaces are mostly present in the form of the farmed desert land. The community organic farm has a variety of home-grown fresh vegetables and fruits aiming to green the desert. Furthermore, these fresh organic products are then sold in the area's local market. The idea of permaculture is also promoted, hoping to raise awareness on the benefits of organic food production for the local community and the surrounding Sinai region. As for El Gouna, the spaces are varying from public, semipublic, semiprivate and private according to the function of the buildings around. The green open public spaces are not well designed and not distributed enough around El Gouna, except for some soft scape around the hotels and the residential areas which act as buffer zone to decrease temperature. Generally, open public places such as squares lack shading devices and soft scape (**Figure 7**).

5.6. Mobility and urban transport

The main idea is to have Basata as a car-free community. The car parking spaces are only on the entrances of the project and not allowed further on. The car parking share is almost 0.3 for each dwelling. Walking and biking are the main mobility means. In Habiba, mobility and urban transport has not really been developed sustainably in the Habiba community. Reliance is mainly on transportation by car and automobiles to get to the area. Inside the eco-lodge area, walking is the main mode of transportation as the area is not very big and so there is no need for any other means. However, the farms and learning centre are a bit far from the eco-lodge, and transportation to them requires an automobile. It should be noted that

Figure 7. Urban farm in Habiba farm, and the urban spaces at EL Gouna.

Figure 8. Bike rents and motorcycle.

the entire Nuweiba area is lacking in this particular aspect, as no means of public transport between the different eco-lodges is provided, and public transportation in the area is not quite developed. In the case of El Gouna, although it is considered a residential and touristic area, the main transportation there are bicycles, tuk-tuk and buses in addition to private cars. It is considered a walkable area; however in summer, it becomes so hard to walk or cycle during the day. The central area of El Gouna is just a pedestrian area and just a touristic place without even any economical place. Walking and biking are the main mobility means, and in El Gouna they used shared bus service to reduce car dependency. According to [32], there are several transport services that were launched in 2014 that combined a petrol engine with an electric motor for cars and shuttles. These allowed the ability for cars to run on electricity alone, all around El Gouna, and so reducing the carbon emissions (**Figure 8**).

5.7. Community involvement

The aim of Basata and Habiba is to create sustainable, environmentally friendly income for the local community. The integration of the local community is an integral part of the organisation and thus provides a lot of support to the local Bedouin community. The locals are provided with several job opportunities through their work in both the eco-lodges and specifically in Habiba in community farms and the Sinai Palm Foundation. In both aspects, they are educated and trained to work in the agriculture and farming work. Furthermore, the educational centre in Basata and the learning centre in Habiba provide after-school education for the local Bedouin children. The children also get involved in the farms as a learning experience to learn the value and importance or producing healthy food and farming. The Habiba community also supports the WOMAD project, a project that aims to empower the local Bedouin women and help them in raising funds for their children's education [33]. In addition to this, all surplus funds that come from the beach eco-lodge are directed towards the expansion and reach of such local projects [33]. In both Habiba and Basata, the local Bedouin women are also provided the opportunity to showcase and sell their handmade accessory products to the visiting tourists and eco-lodge guests. The organic farm's collaborations with NGOs, universities and organisations also expand the role of the place in providing a training hub for sustainable farming and development and an important model for the South Sinai region [33].

In El Gouna, the community part is somehow not considered except for few activities. The playground of the school is being used in summer or night for other community activities which increase the community relations. However, the buildings are not constructed by the residents themselves, which lack the community integration in this part. Thus, labours are hired from the surrounding areas to build this type of construction which they need to be trained before (**Figure 9**).

Figure 9. School's playground is turning into community activity in summer at EL Gouna.

6. Recommendation for planning zero-carbon eco-cities

The aim of eco-cities is to build a viable future for humanity with a healthy planet where the Earth, water and air will continue to support our complex renewable-powered ecosystems.

	Challenge	Action Plan
Mobility & Urban Transport	Direct measures such as energy use, CO2 emissions, air pollution, and traffic noise would be the best indicators of the environmental impact of urban transport.	- Introduce the ideas of e-bikes or e-cars. - Improve the intrinsic environmental performance and efficiency of cars and encourage carpooling. - Facilitate infra-structure for walking and cycling and increasing opportunities to use public transport as much as possible. - Increase opportunities to walk or cycle, or even to participate in activities without moving.
Energy Resources	Water: Scarcity in water. Recycling: Materials are from the earth, keep using them without recycling means lose more of our resources. Electricity: Using energy and fossil cause pollution	-Rise local awareness of water consumption & use water efficient domestic taps. -Use new techniques of recycling water and reuse it, like grey water and black water reuse. - Desalination of sea water using solar energy. It is a fact now that to sustain your resources you must start by recycling current wastes and materials. Invest to use renewable solar, wind or hydro power. Use organic wastes as bio-fuel for public transport.
Architecture & Buildings	Buildings not compatible with energy efficiency codes which consume carbon emission.	Improve building envelope insulation. Apply energy efficient construction methods. Increase efficiency of cooling using passive strategies. Consider vernacular architectural passive solutions after qualitative measurements for verification. Develop energy efficient strategies and applications based on vernacular concepts of sustainability.
Community Involvement	Neglect locals participation in decision making process is one of the current problems that affects the community acceptance of environmental policies.	Encourage local community's participation starting from forming policies to implementation. Encourage the development of design elements that engage residents in small communities in many aspects of natural and human resources management.
Green & Urban space	Cities are suffering from the lack of green urban spaces which caused many problems as pollution that affect the quality of life.	Create more green urban spaces. Encourage carbon offset and reduce carbon foot print
Laws & Legislations	Adoption of traditional building model without any intervention and studying the architectural heritage.	Encourage development of design elements that engage residents in small communities in as many aspects of natural and human resources management as possible.

Table 1. Recommendations and proposed solutions for current challenges to be used as future for guidelines in designing eco-cities in Egypt.

http://dx.doi.org/10.5772/intechopen.80725

In this chapter, it is clear that the concept of eco-community can be achieved apart from the government. So, it is even easier if there is an obligatory design guideline and strict laws and legislations. Here in this section, we are recommending a group of ideas that can be used as guidelines for designing eco-cities based on the analysis of the three case studies. We tried to form the recommendations as current challenges and proposed solution for guidelines. **Table 1** illustrated some essential points that are advised to be taken into consideration while planning an eco-city.

7. Conclusion

The twenty-first century is shaping up to be a traditional era for the humanity who dwells on this Earth. The pressure we are placing on the planet's resources has become increasingly unsustainable. The resulting problems we face, such as water and resource scarcity, increased energy demands and costs, shrinking fossil fuel reserves and a changing climate, have sounded a wake-up call heard round the world. Those who are heading the call and embracing the need for change are finding the necessary solutions and opportunities not only to address this global set of problems but also to advance and improve humanity's relationship with the living world and improve our quality of life. Much of the stress we impose on the Earth is manifested in the way we design, construct and use our built environment; that means buildings and cities must play a vital role in shaping our sustainable future. They are as much representatives of a global approach to our built environment as they are exemplary buildings.

This chapter introduces a brief understanding of eco-community's definitions and is discussing three of the Egyptian pilot projects as an analytical model for a prospect of a zero-carbon city design and planning in Egypt. The dream of the eco-city in Egypt is a city that is a desirable place to live. It is becoming a necessity to develop an integrated model for an eco-city lead by a multidiscipline group of experts including but not limited to renewable energy, agriculture, eco-sanitation, solid waste management, vernacular architecture, traffic and green transportation, economy, social and cultural studies, services, governance and small and micro enterprises. Renewable energy and energy efficiency are a core sector with close implications on other sectors such as agriculture, transportation, housing and services. If these conditions are fulfilled, the community will be able to perform the required actions to produce all needed services and products resulting in an improvement in the overall economic conditions without degrading the surrounding environment. In Egypt, we need a city that is designed around the individual and the family creating a fully integrated neighbourhood orientated around public spaces and civic amenities and a city that encourages the growth of communities and relationships. Sustainability, a vibrant economy, future viability, scientific excellence and a decent life attitude all melt in the same pot towards the same goal of a post-carbon eco-city model for its inhabitants.

We can conclude from the case studies that the concept of low-carbon and low-impact communities can be transformed from an idea to reality and practice. They are not only a product but also a process for sustainable lifestyle. In Egypt unfortunately, we tend to apply replicable outcome-based approaches. It is high time to look for strategies that prioritise participation. We need to deal with complex challenges in our city design approaches. A holistic approach in dealing with low-carbon economic growth should consider the social, economic and ecological reciprocity. More grassroots and community-driven approaches are needed for equity towards low-carbon urban governance. A formal policy support on a national level is important together with national guidance on sustainable communities and community asset transferee. Designing eco-cities is not anymore about only attempting to reduce GHG emissions and energy consumption. We should tackle it from a more holistic view and consider economic justice, behaviour change and wellbeing in addition to community self-management. Some of the recommendations sound like basic aspects in sustainable city planning in modern cities, but Egypt still lacks such basic concepts of sustainable city design.

Acknowledgements

The authors would like to thank Engineer Sherif El Ghamrawy and Engineer Maged Abdel Aziz for providing all necessary information needed for this research work. Special thanks to Dr. Samah El Khateeb and Architect Mamdouh Saker for being essential parts of initiating the idea of this study and their kind contribution.

Author details

Marwa Dabaieh[1]*, Dalya Maguid[2] and Deena El Mahdy[2]

*Address all correspondence to: mada@create.aau.dk

1 Aalborg University, Aalborg, Denmark

2 British University in Egypt, Cairo, Egypt

References

[1] Stern N. A Blueprint for a Safer Planet: How to Manage Climate Change and Create a New Era of Progress and Prosperity. London: Bodley Head; 2009

[2] Heinberg R, Lerch D, editors. The Post Carbon Reader: Managing the 21st Century's Sustainability Crises. Healdsburg, CA: Watershed Media; 2010

[3] Herring H, editor. Living in a Low-Carbon Society in 2050. Basingstoke: Palgrave Macmillan; 2012

[4] General Organization for Physical Planning (GOPP). 2013. National Development Plan for Egypt 2050

[5] World Health Organization (WHO). Global Ambient Air Quality. 2018. http://www.who.int/airpollution/data/cities/en/ [Accessed: 2018]

[6] El Ghorab HK, Shalaby HA. Eco and green cities as new approaches for planning and developing cities in Egypt. Alexandria Engineering Journal. 2016;**55**:495-503

[7] Black A. Pillars, bottom lines, capitals and sustainability: A critical review of the discourses. International Journal of Environmental, Cultural, Economic and Social Sustainability. 2007;**2**(5):107-117

[8] Connelly S. Mapping sustainable development as a contested concept. Local Environment. 2007;**12**(3):259-278

[9] Verma P, Raghubanshi AS. Urban sustainability indicators: Challenges and opportunities. Ecological Indicators. 2018;**93**:282-291

[10] Hamilton K. Where Is the Wealth of Nations? Measuring Capital for the 21st Century. World Bank Publications; 2006

[11] Moldan B, Janoušková S, Hák T. How to understand and measure environmental sustainability: Indicators and targets. Ecological Indicators. 2012;**17**:4-13

[12] Gilbert R, Stevenson D, Girardet H, Stren R. Making Cities Work. Earthscan, London:1996

[13] Soinin K, Birkeland I. Exploring the scientific discourse on cultural sustainability. Geoforum. 2014;**51**:213-233

[14] Register R. Eco-City Berkeley: Building Cities for a Healthy Future. Berkeley: Berkeley Hills Books; 1987

[15] Tang Z, editor. Eco-City and Green Community: The Evolution of Planning Theory and Practice. New York: Nova Science Publishers; 2013

[16] Todd N, Todd J. From Eco-Cities to Living Machines: Principles for Ecological Design. Berkeley, CA: North Atlantic Books; 1994

[17] Lian K, Gunawansa A, Bhullar L. Eco-Cities and Sustainable Cities- Whither? Lien Centre for Social Innovation; 2007

[18] Girardet H. Cities, People, and Planet: Urban Development and Climate Change. 2nd ed. Chichester: Wiley; 2008

[19] Mol APJ, Sonnenfeld DA, Spaargaren G, editors. The Ecological Modernisation Reader: Environmental Reform in Theory and Practice. London: Routledge; 2009

[20] While A. The carbon calculus and transitions in urban politics and urban political theory. In: Bulkeley H, Castan-Broto V, Hodson M, Marvin S, editors. Cities and Low Carbon Transitions. London: Routledge; 2011

[21] Jonas AEG, Gibbs D, While A. The new urban politics as a politics of carbon control. Urban Studies. 2011;**48**:2537-2544

[22] Seyfang G. Community action for sustainable housing: Building a low carbon future. Energy Policy. 2009;**38, 12**:7624-7633

[23] Seyfang G, Smith A. Grassroots innovations for sustainable development: Towards a new research and policy agenda. Environmental Politics. 2007;**16**(4):584-603

[24] Hopkins R. The Transition Handbook: From Oil Dependency to Local Resilience. Totnes: Green Books; 2008

[25] Riffat S, Powell R, Aydin D. Future cities and environmental sustainability. Future Cities and Environment. 2016;**2**:01

[26] Leduc WR, Van Kann FM. Spatial planning based on urban energy harvesting towards productive urban regions. Journal of Cleaner Production. 2013;**39**:180-190

[27] Viljoen A, Bohn K, Howe J. Continuous Productive Urban Landscapes: Designing Urban Agriculture for Sustainable Cities. Architectural Press; 2005

[28] Coles R, Costa S. Food growing in the City: Exploring the productive urban landscape as a new paradigm for inclusive approaches to the design and planning of future open spaces. Landscape and Urban Planning. 2018;**170**

[29] International Development Research Centre. Cities Farming for the Future: Urban Agriculture for Green and Productive Cities. (R. Van Veenhuizen, & E.-U. Agriculture, Eds.) Canada;2006

[30] Mougeot LJA. In: AGROPOLIS, editor. The Social, Political and Environmental Dimensions of Urban Agriculture. London: Earthscan; 2005

[31] Al-Chalabi M. Vertical farming: Skyscraper sustainability? Sustainable Cities and Society. 2015;**18**:74-77

[32] Shawket IM, Ebaid MA. Adopting sustainability in cities; contributing to a better environment. In: The 1st International Conference Towards a Better Quality of Life. El Gouna; 2017

[33] Habiba Organic Farm. 2017. Retrieved June 28, 2018, from Habiba Organic Farm: https://www.habibaorganicfarm.com/

Chapter 3

Buildings in Urban Regeneration

Agnes Schuurmans, Susanne Dyrbøl and
Fanny Guay

Additional information is available at the end of the chapter

http://dx.doi.org/10.5772/intechopen.81803

Abstract

The number of people living in urban areas worldwide is constantly rising. This puts significant pressure on resources in general—leading to an urgent need for the radical improvement of local infrastructure, especially housing, food, water and waste. Managing urban areas is therefore one of the most important development challenges of the twenty-first century. The built environment is a key element in urban living. Significant investments in infrastructure and further development must be made to accommodate the increasing number of people moving to cities, thereby increasing the pressure on available resources and waste generated. Cities also face a huge energetic refurbishment task. These transformations of cities at the same time offer new opportunities: energetic refurbishment could bring multiple benefits to the citizens, and smart material loops can make the circular economy a reality. The design of and the material used for buildings also have a bearing on other important aspects of sustainability and urban life. They affect factors such as health, thermal comfort, acoustic performance and fire resilience. By taking a holistic approach, buildings have the potential to be a part of the solution of today's and tomorrow's challenges through the creation of sustainable cities. Cities have the power to act and make changes happen within their boundaries.

Keywords: energy renovation, urban regeneration, sustainable buildings, fire resilience, city lead, circular economy, resources, waste

1. Introduction

With around 1.5 million people moving to an urban area every week, the pressure on resources continues to increase and will drive the need to radically improve local infrastructures especially for housing, food, water and waste. Already today, cities are aggregators of materials and nutrients, accounting for 60–75% [1] of natural resource consumption, 50% of global waste production, and 60–80% of greenhouse gas emissions [2]. Managing the built environment is the key to securing a sustainable future. In the EU alone, buildings account for 40% of the final energy consumption, about 35% of greenhouse gas emissions, 50% of all extracted materials, 30% of water consumption and 35% of total generated waste [3].

People living in cities typically spend up to 90% of their time indoors—thus creating more demand for energy for heating and cooling of buildings. For households, this can be expensive—with energy expenditure accounting for up to 16% of an average household's total spending.

A key step to the sustainable design of buildings is to reduce the amount of energy needed in buildings to keep a comfortable indoor temperature throughout the entire year. Typically, 80% of the energy used in the building sector results from heating, cooling, hot water and lighting, while the remaining 20% is generated by construction materials, transport and demolition. So, while a low-energy design is essential, it's also important to design in a more comprehensive way and to consider how the building and its materials can be recycled and reused at the end of their life. The right design can make a building a 'material bank' for the future. Requiring deconstruction, selective demolition, sorting, increasing demand for recycled products and developing (digital) platforms for this will bring the circular economy in the built environment to the next level.

Many large cities host challenging neighbourhoods that are often characterised by multiple social problems, inadequate living conditions and bad reputations. These problems often reinforce each other in a vicious circle, causing a negative trend for the area. The buildings forming these areas are often 40–50 years old and were built at a time characterised by different societal ideals. Time has changed and so has our demand from the built environment. Most of these buildings are in dire need of an upgrade—to lower their energy consumption and improve thermal comfort and the health conditions of the people living in them, as well as to improve aesthetics of the neighbourhood. Urban regeneration is important because we need to ensure that our cities, living spaces and our working spaces are fit for the future and enable the citizens to live a sustainable lifestyle. Urban regeneration may be one of the main tasks for our societies to tackle, but it also provides an opportunity to create high-quality, affordable and sustainable buildings if we manage to upscale and replicate the lessons learned from the many pilot cases around the world.

Cities also face new risks due to climate change causing more extreme weather conditions. A higher frequency of disasters like flooding, hurricanes and fires poses new challenges to cities. Cities should be resilient to these risks too, preventing severe human, social and economic consequences. A resilient city will recover more easily after a disaster.

This chapter describes the results of various studies and publications on the role of buildings in cities to meet the challenges described above.

Section 2 explains the results of a recent European study on the challenges and opportunities of energy renovation in a city context, linking it to urban regeneration.

Section 3 goes into sustainable buildings and the challenges and opportunities for cities to get the built environment to become more circular, more resource-efficient and less wasteful. It provides an overview of available studies based on desk research.

Section 4 gives an introduction in generic terms on the theme of resilient buildings and the contribution of buildings to a more resilient city.

2. Energy renovation as a key element in urban regeneration

In June 2018, ROCKWOOL and the Building Performance Institute Europe, BPIE, published a report with a selection of successful urban regeneration projects and an analysis of the key elements in achieving successful urban regeneration [4]. The results of the study and subsequent use of its results in a paper from Copenhagen Economics [5] are described later.

Many large cities host run-down or challenging neighbourhoods, often characterised by multiple social problems, inadequate living conditions and bad reputations. These problems often reinforce each other in a vicious circle, causing a negative trend for the area (**Figure 1**).

Poor houses ➔ energy poverty ➔ social problems ➔ vandalism ➔ bad reputation ➔ lack of investments ➔ poor houses. The challenge to revitalise these areas is amplified by the fact that these neighbourhoods are often physically and socially detached from the rest of the city.

The buildings forming these areas are often 40–50 years old and were built at a time characterised by different societal ideals. During the 1960s and 1970s, there was a huge need for dwellings—resulting in most constructions being built quickly and on the idea of equality. Time has changed and so has our demand from the built environment. Most of these buildings are now in dire need of an upgrade to lower their energy consumption and improve their thermal comfort and the health conditions of the people living in them, as well as to upgrade

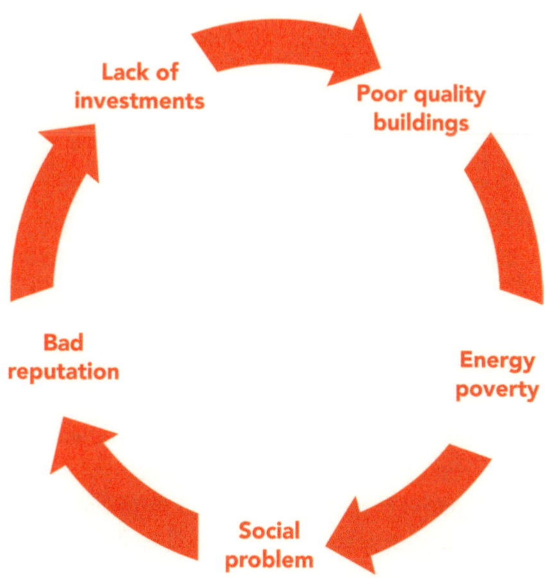

Figure 1. The vicious circle of poor housing in challenging neighbourhoods [4].

the whole neighbourhood that they are part of through the improvement of their aesthetic quality and that of the urban spaces.

Urban regeneration deserves additional attention as cities need to be transformed into living spaces and working spaces in a sustainable manner. Often, challenged urban areas especially need attention in creating resilient buildings as the combination of high-rise buildings and dense urban areas poses a higher risk for the built environment in case of fire or natural disasters. An upgraded building stock will also enable and induce the citizens to live a more sustainable lifestyle.

Many case projects around the world have shown that a comprehensive strategy comprising both physical and social initiatives can transform whole areas into attractive and liveable spaces as well as turn around the negative trend experienced in these neighbourhoods. Upscaling energy renovation of run-down or challenged urban areas is a key element, which can, if combined with other initiatives, not only improve the quality of life and enable citizens to live a sustainable lifestyle but also provide an opportunity to reduce social and health problems in society. Upscaling energy renovation should at the same time focus on creating new business models out of the challenges—focusing on the need for managing resources in the urban environment in a better way.

ROCKWOOL and BPIE describe a selection of successful urban regeneration projects in their study [4]. The key elements in achieving successful urban regeneration summarised later are the result of a review of successful cases combined with an interactive dialogue with leading experts in the field facilitated by ROCKWOOL and BPIE in 2018.

The key learnings identified are especially (see **Figure 2**):

- A strong long-term commitment of and an early engagement from the public authorities is necessary.
- A well-adjusted combination of social and physical measures needs to be considered—which also requires an early collaboration between multiple actors with different expertise.
- The involvement and empowerment of the people living in the area are crucial. Engaging residents in the renovation process, e.g. via 'social contracts', increases the support among residents and can help to give people a new start.

Key success factors which were identified for the building renovation projects were:

- A detailed assessment of renovation alternatives found that a deep renovation would be the cheapest option over a 30-year period.
- Improving the aesthetic quality of the areas combined with the use of long-lasting materials is cheaper in the long run and can attract new investments and new residents to the areas. Doing so can also be a tool to reduce crime and improve quality of life in the area as neighbourhoods that are in a poor state are perceived to be uncared for—resulting in the likelihood of crime going unchallenged [5].
- Building owners see an increased value of their properties while reducing operational and maintenance costs [5].

http://dx.doi.org/10.5772/intechopen.81803

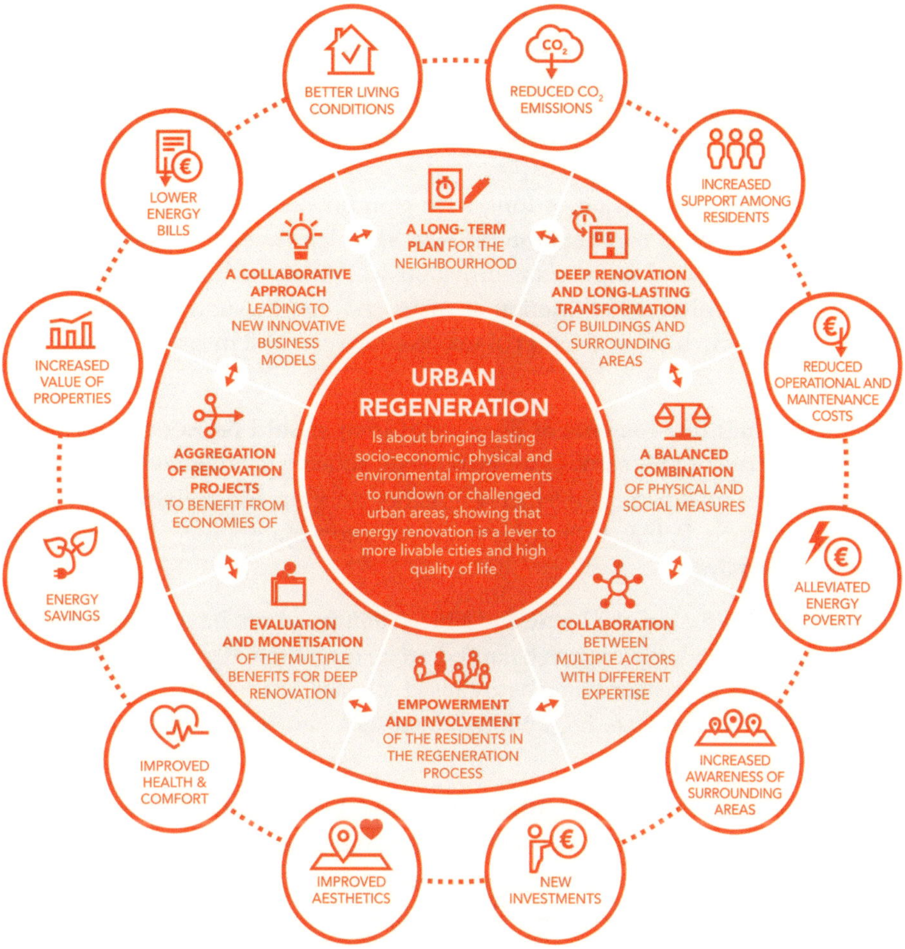

Figure 2. Key elements in achieving successful urban regeneration [4].

2.1. Key elements in achieving successful urban regeneration

One of the key learnings from the many existing urban regeneration projects is that people in general like their homes, want to be informed and involved in large-scale renovation projects and prefer to be given the opportunity to take individual choices.

A deeper dive into the findings provides the following insights.

Urban regeneration is not the same as gentrification. It is, therefore, important to **make sure the residents are involved and empowered in the regeneration process.** While many of the run-down or challenged areas would benefit from hosting a more mixed population, the means to achieve this cannot be to squeeze out certain groups. Extensive energy renovation will actually improve residents' ability to pay rent through lower energy bills. A successful urban regeneration process can therefore lift a whole neighbourhood including the people in it.

A successful urban regeneration strategy cannot be fulfilled without really knowing the needs of the residents. A **bottom-up approach where residents are involved can uncover**

new innovative solutions while giving the residents a sense of ownership of the process, thus increasing their willingness to participate. In other words, an urban regeneration strategy must build on the existing culture rather than building something new. Large-scale renovation projects can be tailored to the existing culture, e.g. by developing a catalogue of individual measures.

Look beyond the near-term period. **A long-term plan for the neighbourhood will increase the chances of a successful regeneration process.** The plan should incorporate the multiple benefits that a physical and social transformation will entail such as a higher living standard and improved health, the alleviation of energy poverty through lower energy bills, an increase in housing prices, the reduction of maintenance cost and the mitigation of the area's climate impact.

Furthermore, **the multiple benefits of deep renovation should be valued and monetized.** Research from Eurofound [6] shows that inadequate housing is linked with numerous societal costs ranging from health care expenses, higher policing and emergency costs to poor academic performance. Public authorities ought to incorporate these costs and benefits in their cost–benefit analysis.

The **physical transformation of a home and a neighbourhood can have a positive impact on people's perception and behaviour.** A balanced urban regeneration process needs to combine both physical and social measures in a harmonised manner. Physical and social measures are not two separate entities but are highly interlinked, as a new home is commonly associated to new beginnings.

The physical transformation of buildings should be long-lasting. It is evident that deep renovations using high-quality materials, combining the energy upgrade with an architectural upgrade, are the cheapest solution in the long run. Deep renovation guarantees a very low energy demand for the next 30 to 50 years. In addition, maintenance and administration costs are reduced as well. Past experience has shown that renovation projects that were quickly done using cheap materials had a short lifespan—thus requiring renovation again. In contrast, high-quality and durable materials guarantee that renovated buildings continue to look nice for many decades.

In all the cases investigated, the cities/municipalities have played a key role. **The authorities have been successful in driving transformation through public-private partnerships,** in which they safeguard the social aspects. Another key success factor is the breaking down of silos within the public governance, for example, by setting up a task group with various departments (energy, climate, social, budget, etc.) and involving solution providers early in the process. Scaling up deep renovation requires more collaboration between various stakeholders, assembling different skills and expertise.

Good planning can use the 'economies of scale' effect when renovating multiple identical buildings. The buildings **built during the 1960s and 1970s are especially suited to a more industrialised renovation approach,** which could reduce the time spent on site and cost. New business models (one-stop shops) and technologies should be supported and further explored.

http://dx.doi.org/10.5772/intechopen.81803

2.2. Moving forward

It is evident that extensive experience exists from individual case studies and within individual stakeholders. The existing experience must be scaled up—following the success factors from existing urban regeneration projects. Cities should be encouraged to develop their own urban regeneration path and to take responsibility for facilitating an early engagement with stakeholders covering all needed competences in a project.

3. Sustainable buildings as contributors to the circular economy in urban regeneration

The benefits of energy renovation in urban regeneration are not limited to energy savings alone. It is broadly recognised to also benefit people and the environment. Good indoor climate, thermal comfort, acoustic performance and daylight, for example, improve the health and wellbeing of inhabitants and positively affect the productivity in schools, hospitals and of workers in general [5]. Taking a life-cycle approach when considering the materials used in buildings can reduce the environmental impact of the buildings in a city. Moreover, valuing the potential for today and the future creates better economic value. Not surprisingly, there is clear trend towards a more sustainable design and renovation of buildings.

Urban centres are particularly impacted by waste from construction and demolition activities (CDW). The Ellen MacArthur Foundation has identified the built environment to be one of the key sectors of structural waste [2]. For example, in Europe, the average office is used only 35–50% of the time, even during working hours. The UN-IRP study called 'The Weight of Cities' [1] shows that the global 'domestic material consumption' (DMC) of raw materials (including sand, gravel, iron ore, coal and wood) is likely to be in the range of 8–17 tonnes per capita per year at 2050, assuming material use per capita will stabilise in developing countries at lower levels than today's developed countries. However, a DMC range of 6–8 tonnes per capita per year has been proposed as an indicative target for sustainable resource consumption [7]. In 2050, urban mining should be the main source of building materials in cities, with implications for storage, logistics and costs, while citizens will increasingly require a non-toxic environment.

Cities are well placed to tackle the resource and waste issues in the construction sector, with their high concentration of resources, capital, data, and talent over a small geographic territory. Closing the loop in the construction sector and improving design would lead to major benefits in sustainability and in quality of life. This could further support urban policymakers in achieving their objectives when it comes to carbon emissions, mobility, indoor air quality [2] and working towards a non-toxic environment. This transition towards a sustainable, low-carbon and resource-efficient economy—called a 'circular economy'—is considered vital to future-proofing cities and improving quality of life for citizens [8]. Our world economy is considered to be only 9.1% circular at the moment [37]. The Circle Economy states that 'Closing the circularity gap serves the higher objective of preventing further and accelerated environmental degradation and social inequality'—especially since housing and infrastructure needs represent the largest resource footprint [37]. They believe that a circular economy approach enables cities to take practical steps to help reduce emissions, create new jobs and strengthen industries and competitiveness, as well as enhance the health and wellbeing of its citizens. Sustainable buildings contribute to this transition.

3.1. Sustainable buildings

There is not one definition for 'sustainable buildings'. However, what all descriptions have in common is that a holistic approach is taken in the design, construction and demolition processes to minimise the buildings' impact on the environment, the occupants and the community and to maintain economic value. When it comes to the environmental aspect of sustainable buildings, the World Green Building Council describes a 'green' building as 'a building that, in its design, construction or operation, reduces or eliminates negative impacts, and can create positive impacts, on our climate and natural environment. Green buildings preserve precious natural resources and improve our quality of life' [9].

There are several features that can make a building 'green' and additional features to make them more sustainable. Environmental features pertain to resources like materials, water and land and energy and carbon, as well as pollution of air, water and soil. Health-related features pertain to indoor climate, acoustic performance and thermal comfort. Additional features could be related to risks like earthquakes and fires—addressing the resources that the building represents.

Several tools and sustainable building rating schemes are in place to assess the sustainability of buildings. Examples of European-wide used commercial rating schemes are BREEAM, DGNB, HQE and LEED, but local schemes exist too. The European Commission recently launched a beta version of a framework for sustainable buildings called Level(s) [10], which describes the basic relevant features for the European housing stock. In the Netherlands [11] and France [12], building legislation is in place to assess environmental building features, and in Germany [13], a similar approach is taken for the Green Public Procurement of buildings.

The key step to a sustainable design is to reduce the amount of energy needed to operate a building. Typically, 80% of the energy used in a building results from heating, cooling, hot water and lighting, while the remaining 20% is used for construction materials, transport and demolition. However, moving towards a more energy-efficient building stock will change this ratio over time—putting more focus on the materials and the 'beyond energy' features of sustainability. Environmental assessments of buildings are therefore based on a life-cycle approach using various environmental indicators.

3.2. Buildings as Material Banks

Considering that a so-called circular economy is vital to future-proofing cities and improving the quality of life of citizens, according to [8], organisations need to stimulate the development of a circular economy by developing concepts for cities to become more circular. The Ellen MacArthur Foundation describes the built environment in 'circular cities' as 'designed in a modular and flexible manner, sourcing healthy materials that improve the life quality of the residents, and minimise virgin material use. It will be built using efficient construction techniques, and will be highly utilised thanks to shared, flexible and modular office spaces and housing. Components of buildings will be maintained and renewed when needed, while buildings will be used where possible to generate, rather than consume, power and food by facilitating closed loops of water, nutrients, materials, and energy, to mimic natural cycles' [2]. This description shows several aspects of buildings in circular cities, centred around the basic assumption that buildings are '*material banks*'.

Today, buildings are often created with only one function in mind, like being a school, an office or a multi-family home. When societal needs or user preferences change, these mono-functional buildings usually become outdated or even obsolete—resulting in a high rate of building vacancy and premature demolition. Those buildings, no longer suitable for use, are considered as waste and are thus treated as such. By applying circularity principles both to the existing building stock and to new buildings, the concept of waste can be eliminated.

This principle is elaborated in the European project Buildings as Material Banks (BAMB) [14]. According to the BAMB vision, the term 'Buildings as Material Banks' refers to a materialised investment: "the building itself is considered as a materialised savings account for material resources, through which building materials, products and components are temporarily 'deposited' into a functional element or part of the building. When socio-economic conditions are favourable, (a part of) the materials, products and components may be retrieved for another investment, that is: another building or another high-quality application. Seeing material resources as a temporary way of materialising investments opens the door to a wide range of circular business models, in which economic and environmental value is conserved and created through the reuse of materials, products, components and buildings, while (performance-based) services are provided to support the daily life of (end) users."

Three major systemic changes have been identified to support the BAMB vision:

- Change in design culture
- Change in value definition
- Change in collaboration across all actors

Both the existing building stock and the new buildings are interlinked in urban regeneration.

3.3. Resources from the existing building stock

The growing demand for and the scarcity of resources, less space for landfill and the challenges regarding the movement of resources through the city all create a push towards an approach of using the materials and products in existing buildings in the most optimal way. 'Urban mining' is mentioned in several sources as a necessary approach in circular cities [1, 15]. At the same time, it is identified that there are insufficient economic and regulatory incentives as well as a lack of trust in the quality and availability of secondary materials to create a market for recycled building materials [15, 16]. A lack of demand for reused or recycled products in combination with easy and cheap waste landfill and incineration hamper a sound business model throughout the construction chain.

The EU Construction and Demolition Waste Protocol [17] lists seven features for successful implementation of recycling:

1. Ban landfilling of construction and demolition waste.
2. Implement mandatory pre-demolition waste audits (identify quality and quantity).
3. Enforce traceability of waste to establish confidence (especially during transport).
4. Urban planning (recycling facilities within city limits).
5. Manage quality of recycled waste (same quality standards as virgin materials).
6. Environmental management along the entire value chain.
7. Create open markets through leading example of public procurement.

Below, possible measures are discussed to keep material resources from buildings in the loop, for both the end of the building's life and the production of new products.

3.3.1. Measures at the end of the life of a building

Upon demolition, budget constraints do not allow for (often more labour-intensive) the deconstruction of separate material fractions. The limited space for collection of many fractions is a further barrier in cities—requiring a more sophisticated after-sorting. Toxic legacy materials require specific attention if products must be reused or recycled. In many countries, landfill or incineration are still easily accessible and are too cheap to incentivise economically feasible alternatives. Separation and recycling only work business-wise in countries or regions with strict landfill bans for recyclable materials—like in Denmark or the Netherlands. This must be combined with (legal) requirements or incentives for buildings to be stripped before their frames can be demolished following an assessment (pre-demolition audit). This can create an entire service sector of SMEs with expertise in the removal of tiles, flooring, window frames, insulation material, sanitary items, lighting, etc. The Rotor in Flanders, Belgium, is an example of such a specialised company, as presented on the event from GLOBE-EU [15].

Cities can lead the way by discouraging landfill and incineration of construction and demolition waste and stimulating deconstruction, sorting and facilitate reuse platforms like Rotor. Local urban initiatives also contribute to social cohesion and job creation [8].

To facilitate this further, it is suggested to open up the European concept of waste [15, 18], in which 'preparing for reuse and recycling' is part of the waste definition, as shown in the left-hand side of **Figure 3**. On the right-hand side, an additional category is added between non-waste and waste (refuse), representing a situation in which secondary resources for reuse and recycling will not be regarded as waste to better fit the circular economy. Cities will have much more room for policies and experiments, if the European Commission would issue guidelines on the conditions to be fulfilled for recovered or reclaimed materials

http://dx.doi.org/10.5772/intechopen.81803

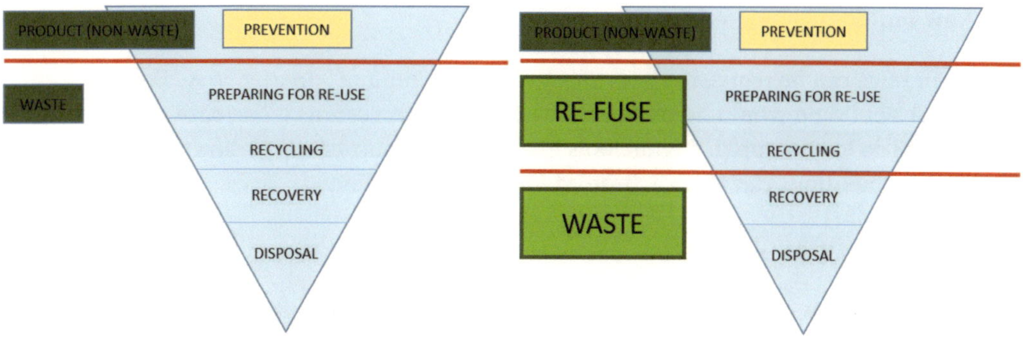

Figure 3. An illustration of the application of current definitions of waste [18].

to be considered a 'non-waste' while guaranteeing current levels of protection of public health and the environment. It could be similar to the conditions for by-products of manufacturing processes [15], so that materials, such as bricks, fittings, doors, window panes, beams, etc., can be transported, stored, processed, and sold as products provided these conditions are met.

3.3.2. Measures to stimulate the use of secondary materials

Closing the route towards landfill and incineration must go together with boosting demand for reused products and secondary materials. The market for secondary raw materials is still immature, and several policy choices exist to boost this market. The policy toolkit report from the Ellen MacArthur Foundation [38] explains it as follows: 'There are two broad, complementary policymaking strategies that can help accelerate the circular economy. The first is to focus on fixing market and regulatory failures. The second is to actively stimulate market activity by, for example, setting targets, changing public procurement policy, creating collaboration platforms and providing financial or technical support to businesses. These approaches are complementary and policymakers can determine where to put the emphasis, taking inspiration from the most applicable aspects of both approaches'.

Within public procurement, cities have the power to set high thresholds for recycled content in their public tenders and encourage the use of reusable and recyclable materials and products. Other suggestions are fast-tracked approval of construction permits, easier access to capital and stimulating new ownership models of buildings and constructions [15]. Creating platforms for collaboration and promoting value networks instead of value chains could also be a role for cities, as shown by the city of Brussels in the BAMB project [14] and the Dutch city of Almere in their UpCycle City contest [19]. A key takeaway from the Amsterdam Circular is to work closely together with the private sector and research institutes to find new business models in the context of existing strategies, such as green procurement.

Note that there is a growing belief that markets should be created without government interference like subsidies and levies on raw materials but much more by facilitating the emergence of business models that can work over time.

3.4. New buildings prepared for the future

New buildings can be requested to be designed and built in a 'circular' way. Better information about a building would help close the loop, and lower costs as worst-case scenarios in the future need no longer apply. Digital tools like BIM and cloud-based platforms of comprehensive data on buildings can already be used for today's buildings.

3.4.1. Reversible design and open building concept

Modularity and flexibility are mentioned by the Ellen MacArthur Foundation as key to circular buildings in cities [2]. A Dutch Architect Manifesto refers to both adaptability/flexibility and easy-to-dismantle construction [20]. Research on such reversible building designs is broadly available, as shown in the BAMB project [14] and the EU-based project [21], although it is not a common practice yet. The Flanders region in Belgium has applied these principles in their policy for 'Building for change', a policy that anticipates on future changes in building use, as presented in [15].

An 'open building' concept is based on similar principles, thereby taking the perspective of ownership. This seems especially interesting in the urban environment, with many rental buildings and apartment blocks. This concept suggests that buildings need to be looked at from the perspective of the owner (long-term client) and the tenant (short-term client), respectively. The diverging interests of these two owners make it important to treat them separately (two-step housing). Investors in real estate are interested in the 'shell' of a building, not the interior outfit, which is considered more of a liability. Cities in Japan are more advanced in the implementation of an open building concept which recognises that people need to adapt their living space according to changing circumstances. An entire industry sector has emerged around the remodelling of homes based on circularity principles: easy-to-remodel interiors, reuse of components and floor space tailored to family size [15, 22].

Encouraging innovation by adding reversible design to the conditions for competitive bidding on public works contracts and the development of EU-wide indicators and standards for reversible building design would go a long way towards more sustainable building design and use.

3.4.2. Information throughout the chain

Information on the materials in a building and on the building itself over time facilitates a building to be a material bank in future. The digitalisation of society will support this.

Material passports are based on the principle that one should know what type of materials is applied in the building. Information on content should reveal potential future legacies of hazardous materials that could hamper recycling. Further information, e.g. on recyclability, is suggested [14, 23]. It is likely that building information modelling (BIM) will also play a role in the transfer of environmental product information through the chain, as it already fulfils this role for technical characteristics.

Whereas material passports are typically connected to products and their manufacturers, building passports could play a role in city policies. Building passports provide information on the whole building design, its maintenance, refurbishments, etc. Building passports are not well defined yet and are suggested for different purposes, including renovation and circular economy, but they could also be a helpful tool for risk evaluation in case of, e.g. a fire or natural disaster. An example of a building passport in the context of the circular economy is Madaster [24]. It is Madaster's mission to eliminate waste by providing materials with an identity. The Madaster Platform facilitates the registration, organisation, storage and exchange of data of buildings over their lifespan, thereby becoming a public, online library of materials. The concept of building passports needs further development to be used in a consistent way on a large scale, but the principle clearly has a place when developing more circular buildings in an urban context.

3.4.3. What is a circular building?

It could be helpful in procurement and design processes if a definition of 'circular buildings' would exist. This is not the case, as the possibility of having a singular definition is being challenged, and the idea that 'circular buildings' are multidimensional is being considered.

Several suggestions have been made, for example in [2, 14, 20, 25]. Common themes in the descriptions of 'circular buildings' are energy efficiency, the materials applied in the building (their use of resources, reuse and recycling options and the waste they cause) and a design that enables long-term benefits from the building and its components.

Such a 'circular building' is not similar to a 'sustainable building': sustainability is to be assessed, but a circular building puts more focus on design and saving resources while not compromising sustainability. For the environmental assessment, an LCA approach as described before is indispensable. LCA reveals any trade-offs of specific measures targeting specific issues, like resources or waste in the circular economy. In this regard, it is important to remember that the circular economy and circular buildings target a goal—usually of reducing resource consumption and waste—*but 'circularity' is not a goal in itself!* Not all circular concepts will result in environmental benefits in the end. It can be questioned, for example, if materials should be recycled into products that are not recyclable anymore—a pitfall for some innovative products made from waste.

4. Resilient buildings

4.1. Resilience

The term resilience has been around for centuries and has been used in engineering, ecology and psychology before coming to civil protection and disaster risk reduction.

Several definitions of the concept of resilience can be found in the literature, for example, in publications from Timmerman [26] and Holling [27]. The UN Office for Disaster Risk Reduction came with the following definition in 2009:

> *"The ability of a system, community or society exposed to hazards to resist, absorb, accommodate, adapt to, transform and recover from the effects of a hazard in a timely and efficient manner, including through the preservation and restoration of its essential basic structures and functions through risk management." [28].*

Today, increased risk to hazards is manifested in cities. Climate change (flooding), earthquakes and terrorism, for example, pose new challenges to cities. Cities should be resilient to prevent social breakdown, physical collapse or economic deprivation. The Rockefeller Foundation report for the City Resilience Framework [29] describes city resilience as 'the capacity of cities to function, so that the people living and working in cities – particularly the poor and vulnerable – survive and thrive no matter what stresses or shocks they encounter'. It is stated that risk assessments and measures to reduce specific foreseeable risks will continue to play an important role in urban planning. Buildings also have an important role to play in this area. In land-scarce urban areas, more and more people are working and making their homes in high-rise buildings, which impose a challenge on safety issues if not well taken care of. If an earthquake or a fire strikes in a dense city, the consequences can be very serious. Energy efficiency and sustainability should—and can—go hand in hand with more resilient buildings.

4.2. Resilient and sustainable buildings

In resilient cities, man-made infrastructure and buildings are well-conceived, well-constructed and safeguarded against known hazards. Building codes and standards should promote long-term robustness, flexibility to adapt in the future and safe failure mechanisms in the event of a shock [28].

It has been argued that design for resilience is to design for sustainability to reduce the environmental impacts and societal consequences of post-hazard repairs [30] (as referenced in [31]). Conversely, design for sustainability is to design for resiliency to prevent that the unlikely extreme events may impact the urban communities [31]. This study shows such an integrated approach for a holistic building design, considering safety, resiliency and sustainability, based on multiple conflicting criteria. The US building assessment scheme RELI has its own system of not only assessing all types of risks related to resilient buildings but also starting from the sustainable building perspective [32].

It should be realised that building codes and standards often provide *minimum* requirements, focusing on safe escape of people. Standard EN 16309 for assessing sustainable constructions starts from this principle, by valuing 'above codes' performance as being more sustainable [33]. Accidental actions (earthquake, explosions, fire, traffic) are sustainable building aspects to assess. This shows once more the interlinkage of resilient and sustainable buildings.

The new EU Energy Performance of Buildings Directive (EPBD) from 2018 may lead to some improvements towards holistic building design, taking both energy performance and hazards into account. It requires for renovation projects to address fire safety and risks related to intense seismic activity affecting energy efficiency renovations and the lifetime of buildings, as well as the issue of healthy indoor climate conditions (Art.2a § 7 and Art.7 § 5).

4.3. Fire-resilient buildings

The above section clarifies that fire resilience also has a connection to sustainable buildings. The fire resilience of a building is often described as the ability of a building to withstand the effects of a fire, or it is often linked to the building's fire resistance properties. However, fire resilience covers more than the technical characteristic of being fire-resistant—it considers how the environment, the community and economy adapt and recover from a fire.

Fire can cause disturbance to a city's ability to deliver its service to the community. Especially in densely populated urban areas, the built environment is a crucial part of the city's infrastructure. In January 2018, a fire in a hospital in South Korea killed 37 people and injured over 100. All patients had to be evacuated and the hospital was immediately closed. This meant that this hospital was no longer able to serve the community, and the remaining patients had to be relocated somewhere else. To improve resilience, it is recommended to include property protection of critical infrastructure, in addition to the current requirements for life safety in case of fire [34].

Also, efforts to improve the sustainability of buildings often focus on increasing energy efficiency and reducing the embodied carbon. However, a fire could reduce the overall sustainability of a building through the release of pollutants and the subsequent rebuild. Furthermore, fires have a range of less immediate and obvious adverse consequences on the natural environment. These include air contamination (which is likely to also include land and water contamination), contamination from water runoff containing toxic products and other environmental releases from burned materials.

As stated earlier, there is a disconnection between sustainable building approaches and building regulations. When designing a sustainable and energy-efficient building, architects usually focus on elements such as using solar energy and promoting air circulation. In parallel, fire safety requires the respect of building codes and other specific national requirements focusing on other elements such as safe escape routes and optimal operating conditions for fire fighters.

Moreover, sustainable building assessments are based on a life-cycle description of buildings, which is scenario-based—scenarios that do not plan for fire. However, statistics show the scenario of a fire occurring during the lifetime of a building is a realistic scenario altogether. In the USA, for example, an estimated 380,200 residential building fires were reported to fire departments each year between 2013 and 2015 [35].

Therefore, methods that include quantitative risk assessment (QRA) for predicting fire spread to adjacent structures; life-cycle assessment (LCA) for estimating the environmental impact, the fire response and replacement of damaged materials; and cost–benefit analysis (CBA) for estimating the economic impact of the fire have been proposed to analyse optimal designs and environmental consequences [36]. These methods could be one solution to include fire resilience in sustainable building assessment.

In summary, fire safety, sustainability and energy efficiency should be assessed together. A risk assessment should be performed both when building new and when renovating as fire does not only have an impact on the structure of a building but also has societal, environmental and economic impacts that we should try to mitigate.

5. Conclusions

Managing urban areas and resources is one of the most important development challenges of the twenty-first century. Moreover, efficient and resilient buildings play a major role as part of the urban infrastructure.

The development in European legislation of buildings and several initiatives of cities themselves are going in the direction of a more holistic and integrated approach that includes accidental hazards like fire as an integral part of sustainable buildings and energy renovation approaches.

With an increasing demand for buildings in cities and an ageing and outdated building stock, there is a huge potential for urban regeneration to use the built environment as a key element towards achieving both environmental and social objectives.

Buildings have the potential to be part of the solution of today's and tomorrow's challenges in sustainable cities by taking a holistic approach. Extensive experience already exists across the world for how to build and renovate in a sustainable way to achieve high-quality, low-energy-consumption and resilient buildings. The main task is to find ways to scale the existing activities and to develop new and viable business models around the challenges faced in managing urban areas and resources.

Author details

Agnes Schuurmans*, Susanne Dyrbøl and Fanny Guay

*Address all correspondence to: agnes.schuurmans@rockwool.com

ROCKWOOL Group, Hedehusene, Denmark

References

[1] UN Environment - IRP. Swilling M, Hajer M, Baynes T, Bergesen J, Labbé F, Musango JK, et al. The Weight of Cities: Resource Requirements of Future Urbanization. Report by the International Resource Panel. United Nations Environment Programme, Nairobi, Kenya. 2018. Available from: https://citiesipcc.org/wp-content/uploads/2018/03/Summary-for-Policymakers_Global-Weight-of-Cities-1.pdf [Accessed: Jun 30, 2018]

[2] Ellen MacArthur Foundation. Cities in the Circular Economy: An Initial Exploration. 2017. Available from: https://www.ellenmacarthurfoundation.org/assets/downloads/publications/Cities-in-the-CE_An-Initial-Exploration.pdf [Accessed: Jun 30, 2018]

[3] European Commission (DG ENV), Biois, Arcadis, IEEP, A project under the Framework contract ENV.G.4/FRA/2008/0112 Service contract on management of construction and demolition waste – SR1, Final Report Task 2, February 2011. Available from: http://ec.europa.eu/environment/waste/pdf/2011_CDW_Report.pdf [Accessed: Jun 30, 2018] As referenced in the communication from the commission to the european parliament, the council, the European economic and social committee and the committee of the regions on resource efficiency opportunities in the building sector (2014) 445 final, p.2. Available from: http://ec.europa.eu/transparency/regdoc/rep/1/2014/EN/1-2014-445-EN-F1-1.Pdf [Accessed: Jun 30, 2018]

[4] ROCKWOOL, BPIE. Upscaling Urban Regeneration—European frontrunner cases are leading the way. 2018. Available from: http://www.ROCKWOOLgroup.com [Accessed: Sept 18, 2018]

[5] ROCKWOOL, Copenhagen Economics. Putting Renovation on the Agenda—Global perspectives of the value of renovation. Available from: www.ROCKWOOLgroup.com [Accessed: Sept 26, 2018]

[6] European Foundation for the Improvement of Living and Working Conditions, Inadequate Housing in Europe: Costs and Consequences. Available from: https://www.eurofound.europa.eu/sites/default/files/ef_publication/field_ef_document/ef1604en_0.pdf [Accessed: Jul 2, 2018]

[7] Fischer-Kowalski M, Swilling M, von Weizsäcker EU, Ren Y, Moriguchi Y, Crane W, et al. UNEP. Decoupling natural resource use and environmental impacts from economic growth. A Report of the Working Group on Decoupling to the International Resource Panel. 2011. ISBN: 978-92-807-3167-5

[8] Towards Circular Cities. 2017. Available from: www.eurocities.eu [Accessed: Feb 17, 2018]

[9] World Green Building Council. 2016-2018. Available from: http://www.worldgbc.org/what-greenbuilding [Accessed: Jul 5, 2018]

[10] European Commission. Level(s) Building Sustainability Performance. 2018. Available from: http://susproc.jrc.ec.europa.eu/Efficient_Buildings/docs/170816_Levels_EU_framework_of_building_indicators_Parts.pdf [Accessed: Jun 14, 2018]

[11] Stichting Bouwkwaliteit. Assessment Method Environmental Performance Construction and Civil Engineering Works. November 2014. Available from: https://www.milieudatabase.nl/imgcms/SBK_Assessment_method_version_2_0_TIC_versie.pdf [Accessed: Nov 12, 2015]. As referred to in Article 5.9 of the Dutch Building Decree 2012. Available from: https://www.bouwbesluitonline.nl/Inhoud/docs/wet/bb2012/hfd5 [Accessed: Jun 2, 2018]

[12] E+C− Building Regulation France. 2018. Available from: http://www.batiment-energiecarbone.fr/en/home/ [Accessed: Jul 5, 2018]

[13] UBA. July 23, 2010. Available from: https://www.nachhaltigesbauen.de/anerkannte-systeme-in-deutschland.html [Accessed: Jul 6, 2018]

[14] BAMB. Buildings as Material Banks. 2015. Available from: https://www.bamb2020.eu/ [Accessed: Jun 21, 2018]

[15] GLOBE EU. Recommendations from "Re-using the Rubble" Green Week 2018 Pre-conference, May 22, 2018. Available from: http://www.globe-eu.org/wp-content/uploads/GLOBE-EU-Recommendations_22052018.pdf and http://www.globe-eu.org/wp-content/uploads/GLOBE-EU-Background-Paper_22052018-.pdf [Accessed: Jun 4, 2018]

[16] European Commission, Com (2015) 614/2, Communication from the Commission to the European Parliament, the Council, the European Economic and Social Committee and

the Committee of the Regions. Closing the Loop—An EU Action Plan for the Circular Economy. Available from: https://ec.europa.eu/transparency/regdoc/rep/1/2015/EN/1-2015-614-EN-F1-1.PDF [Accessed: Jun 2, 2018]

[17] EU Construction and Demolition Waste Protocol. Sept 18, 2018. Available from: https://ec.europa.eu/growth/content/eu-construction-and-demolition-waste-protocol-0_el [Accessed: Jul 4,2018]

[18] Eurocities. Draft Action Plan 09.02.2018. Urban Agenda for the EU, Circular Economy. Available from: www.eurocities.eu [Accessed: Feb 17, 2018]

[19] Eurocities. Brochure Full Circle. Cities and the Circular Economy. Oct 2017. Available from: www.eurocities.eu [Accessed: Feb 17, 2018]

[20] BNA. Manifest Circulaire Architectuur. 2018. Available from: https://www.bna.nl/programmas/duurzame-ontwikkeling/wij-gaan-circulair/manifest-circulaire-architectuur/ [Accessed: Jan 31, 2018]

[21] Conception Construction Zéro Déchet. 2012. Available from: http://www.bazed.fr/ [Accessed: Nov 30, 2017]

[22] Zuidema R. The Elephant Rumbles through the Circular City. Jan 31, 2018. Available from: https://www.briqs.org/ [Accessed: Mar 21, 2018]

[23] Luscuere LM. EPEA Nederland, materials passports: Optimising value recovery from materials. In: Proceedings of the Institution of Civil Engineers Waste and Resource Management 170, February 2017 Issue WR1, ICE Proceedings. pp. 25-28. DOI: 10.1680/jwarm.16.00016 (Paper 1600016)

[24] Madaster. 2018. Available from: https://www.madaster.com/en [Accessed: Jul 4, 2018]

[25] WE Consultants. Circulair Bouwen is Meetbaar. Blog pril 12, 2017. Available from: https://www.w-e.nl/circulair-bouwen-is-meetbaar/ [Accessed: Jul 2, 2018]

[26] Timmerman P. Vulnerability, resilience and the collapse of society: A review of models and possible climatic applications. Environmental Monograph. 1981;**1**:1-45. DOI: 10.1002/joc.3370010412 [Accessed: Jul 3, 2018]

[27] Holling CS. Resilience and stability of ecological systems. Annual Review of Ecology and Systematics. 1973;**4**:1-23

[28] UNISDR. Terminology. 2017. Available from: http://www.unisdr.org/we/inform/terminology [Accessed: Jul 2, 2018]

[29] Arup, the Rockefeller Foundation. City Resilience Framework. 2014 (Updated December 2015). Available from: https://assets.rockefellerfoundation.org/app/uploads/20140410162455/City-Resilience-Framework-2015.pdf [Accessed: Jul 5, 2018]

[30] Wei H, Shohet IM, Skibniewski M, et al. Assessing the lifecycle sustainability costs and benefits of seismic mitigations designs for buildings. Journal of Architectural Engineering. 2016. p. 22

[31] Alibrandi U, Mosalam KM. A Decision Support Tool for Sustainable and Resilient Building Design. Germany: Springer;

[32] Resilient Building Design—RELi. 2014. Available from: http://c3livingdesign.org/?page_id=13783 [Accessed: Jul 5, 2018]

[33] EN 16309:2014. Sustainability of construction works—Assessment of social performance of buildings—Calculation methodology. Available from: https://standards.cen.eu/dyn/www/f?p=204%3A110%3A0%3A%3A%3A%3AFSP_PROJECT%2CFSP_ORG_ID%3A58840%2C481830&cs=1EB6AC88BE9625D8C607EF660D11FEEB3 [Accessed: Jun 2, 2018]

[34] Gernay T et al. Urban infrastructure resilience to fire disaster: An overview. In: World Multidisciplinary Civil Engineering-Architecture-Urban Planning Symposium 2016, WMCAUS 2016. Procedia Engineering: Amsterdam, The Netherlands: Elsevier; 2016. pp. 1801-1805. Available from: https://www.sciencedirect.com/ [Accessed: Jul 4, 2018]

[35] FEMA, US Fire Administration, Residential Building Fires (2013-2015), Topical Fire Report Series. June 2017;**18**(1). Available from: https://www.usfa.fema.gov/downloads/pdf/statistics/v18i1.pdf [Accessed: Jul 5, 2018]

[36] Amon F et al. Development of an Environmental and Economic Assessment Tool (Enveco Tool) for Fire Events. Final Report. SP Technical Research Institute of Sweden; New York: Springer; 2016

[37] CircleEconomy. The Circularity GAP Report. 2018. Available from: https://www.circularity-gap.world/ [Accessed: Sep 18, 2018]

[38] Ellen MacArthur foundation. Delivering The Circular Economy. A Toolkit for Policymakers. 2015. p. 32. Available from: https://www.ellenmacarthurfoundation.org/assets/downloads/publications/EllenMacArthurFoundation_PolicymakerToolkit.pdf [Accessed: Sep 18, 2018]

Chapter 4

Urban Renovation and the Simulation Evaluation of Urban Climate Change in Residential and Commercial Districts: A Case of Xi'an, China

Wang Yupeng, Ma Dixuan, Li Man and Zhou Dian

Additional information is available at the end of the chapter

http://dx.doi.org/10.5772/intechopen.79267

Abstract

The urban heat island (UHI) effect has drawn attention to monitor and evaluate outdoor thermal comfort in cities worldwide. The rapid, large-scale urban development in China is producing urban climate change in large cities, creating other urban environmental problems such as haze weather, which is one of the most important environmental issues in China. High-density building development will change the urban typology, leading to changes in the urban sky view factor (SVF) and microclimate. Since the energy consumed by indoor heating and air conditioning is highly related to the outdoor mean air temperature, a high SVF should be considered in the planning period. In this chapter, the typical urban planning styles in China are evaluated. Four microscaled residential and three commercial districts in Xi'an city are selected, to represent the typical urban typology of residential and commercial districts that developed during different historical periods and used the urban simulation system scSTREAM to evaluate the impact of urban renovation types on urban climate change.

Keywords: urban renovation, urban typology, climate change, outdoor comfort, environmental simulation

1. Introduction

1.1. Urban climate change in cities

In recent years, urban climate change has been observed in most of the world's developed cities. In central Beijing, the annual air temperature rose 0.36°C during 1961–1980. However, during the building boom between 1981 and 2000, it rose 0.94°C [1]. The correlation coefficient between the impervious surface rate and land surface temperature in Beijing reached 0.93, which means the impervious of urban surface is providing a large contribution to urban climate change [2]. Climate models indicate that due to the expected warming of up to 9°C by the 2080s in the Arctic and the southern and central Prairies [3], the number of days with average temperatures above 30°C is likely to increase in cities across Canada, especially those in the Windsor-Quebec corridor (such as Toronto) and portions of British Columbia. Thus, urbanization patterns, especially in the central parts of cities, have a large impact on urban climate change. The spatial variability of urban heat islands (UHIs) in cities has been found to be a function of urban surface properties, which in turn are influenced by land cover, especially vegetation cover and building density [4].

The deep urban canopy created by high rises can increase the wind speed in urban areas and affect the urban thermal environment. A simulation comparison of high-rise and low-rise buildings in the Lujiazui district of Shanghai found that with low-rise buildings, wind speed declined 22%, air temperature decreased 7%, and O^3 decreased 9% [5]. Another study used wind tunnel measurements to examine wind velocities in Toronto, confirming that among several high-rise towers, wind often accelerated above 10 m/s; this created wind-chill effects and exerted mechanical forces on pedestrians, making it unsafe for them to walk [6]. The openness of urban geometry can be defined using the sky view factor (SVF). The correlation between SVF and the urban thermal environment has been demonstrated in Montreal, Canada [7–9]. A high SVF, which means more open urban space, could be related to a lower UHI index.

Urban development largely serves the purposes of economic development. Especially in China, large-scale, rapid urban development mostly focuses on the operational efficiency of cities, with little attention to the long-term environmental effects. This is a major cause of China's current environmental crisis, and the problem is rapidly spreading to India as well as other countries of Southeast Asia and the Middle East.

1.2. The urban typology changes of cities

With the accelerated speed of urban development and the constant expansion in Chinese cities, the urban typology has undergone drastic changes in the past 10 years. Xi'an is a historical city as well as one of the most developed cities in China's central plains, which is experiencing rapid urbanization development, urban renewal and expansion. Most of the current urban buildings were constructed after 1979 [10]. Different development styles and residential building types can be observed for the different periods of rapid urban development. The urban typology changes of Xi'an is the epitome and representative that of China.

In China, the government started to pay attention on economic development and infrastructure redevelopment after 1979. In the process of redeveloping existing urban areas, new construction was continuously built on the edges of central urban areas. In order to meet the residential demand for citizens, large factories led to developed residential communities around the factories for their staff and workers. During this period, most of the buildings are built in 5–6 stories, and some of these have seven floors.

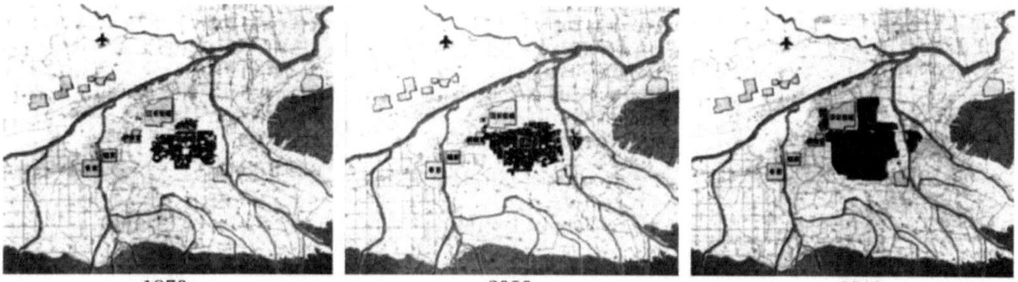

Figure 1. Urban expansion in Xi'an [11].

After the high-speed urban expansion, real estate developers started to have a major impact on urban reform after 1990s. New large-scale developments started occurring outside the city core. During this period, the sense of building design and construction quality became more important than that in the past. High-rise and detached houses were also developed in some of the projects, and most buildings were still 5–7 stories.

After 2000, because of the increased population density, the high-rise building has become the most common construction style in residential development projects. The first demand of residential construction has shifted from meeting the citizens' living to promoting real estate and the urban economy status. The process of urban expansion in Xi'an is showed in **Figure 1**. It is clearly showing that urban occupation has grown rapidly during 2000–2010. Meanwhile, urban building density has also increased because of the high building density of recent development projects.

2. Methods

Residential building is the most typical form of architecture in the city as some related research points out. From 2000 to 2010, area of residential land accounts for the largest proportion in urban land use type [12]. In this study, typical urban planning styles in China were selected and analyzed. Microscale residential districts in Xi'an were selected for representing the typical urban typology of residential districts that developed during different periods and used the urban simulation system scSTREAM (Software Cradle Co., 2011) to evaluate the impact of urban typology change on urban climate change.

2.1. Urban typology in Xi'an city

To estimate the impact of urban typology to urban ventilation and urban air quality, four typical residential areas were selected. The first area, Sanxuejie, is a low-rise area, a traditional residential zone rebuilt after the 1950s. It retains the urban form used in China Science the Ming and Qing Dynasties (about 600 years ago), with narrow streets between residential buildings, most of which have two floors. The second area, Xitiedaminggong, is a middle-rise neighborhood developed in the early 1990s. All of the residential buildings are five floors in height and perfectly represent the character of the building type constructed in the period between 1979 and the beginning of the 1990s. The third neighborhood, Jiaodayicun, is a mix-rise area of low-rise and high-rise buildings, home to a variety of building styles developed around the 1980s and 2000s. The last selected neighborhood, Gongyuantianxia, is a high-rise residential neighborhood developed in 2009. The average building floor in this area was 17 floors in height, representing the common development style after 2000 (**Figure 2**).

Besides, three typical shopping areas in different layout which present the typical urban typology of shopping district are selected. A domain of 500 × 500 m was chosen for the simulation models, and the detailed input domain data and weather data are presented in **Tables 3** and **4**.

The first is Xiao Zhai shopping district, which is located in Xi 'an Yan Ta district. Since 2001, it has become the second largest business circle in Xi 'an, and the area is mainly composed of small retail shops and large shopping centers, and these commercial buildings have no unified planning, and they are not built in the same period. In this district, they also have some residential architectures. Therefore, it can be said to be a traditional kind of mixed layout shopping district, and it is common in Xi'an. The second is Shu Yuan men shopping district, and it is located in the east side of the south gate in Xi'an. It has been developed since the Ming dynasty, and it has become a kind of antique commercial street with Ming dynasty and Qing dynasty architecture style. In this area, building density is relatively high, but building height is low, generally only 2–3 levels of height. Meanwhile, there are some old residential one-storied houses in the area. Therefore, it is a traditional historical and cultural district for shopping. The third is Tang West Market Group shopping district. It is the only project that was rebuilt on its original site in China. Through overall planning and design, this area was

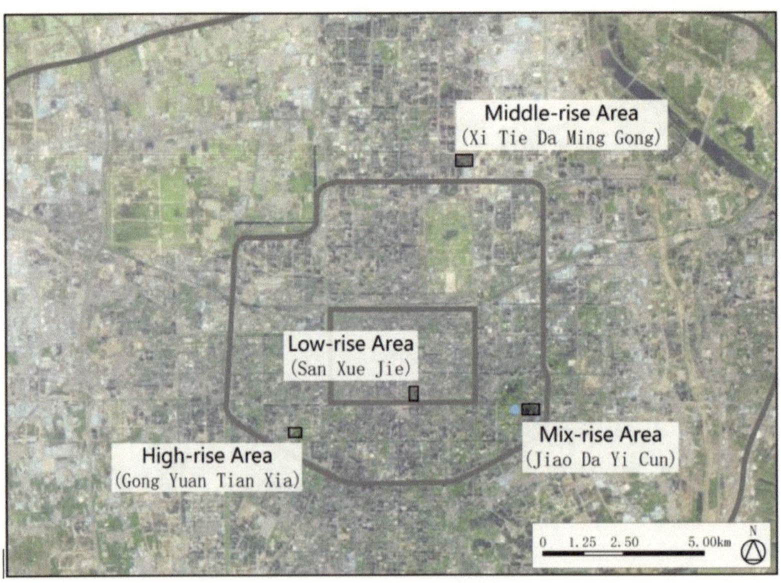

Figure 2. Xi'an satellite photo with the four areas selected for this study and defined as low-rise area (Sanxuejie), middle-rise area (Xitiedaminggong), mix-rise area (Jiaodayicun) and high-rise area (Gongyuantianxia). The area represented in this figure is indicated in **Figure 3** and **Tables 1** and **2**.

open in 2012. In this area, there are one big shopping center, one supermarket, one hotel, one museum and an antique market. They are basically multistoried buildings. Therefore, it is a new mixed-use shopping district.

	Low-rise	Middle-rise	Mix-rise	High-rise
Site size (m)	280 × 440	520 × 364	450 × 340	410 × 300
Simulation size (m)	1080 × 1240	1520 × 1364	1450 × 1340	1410 × 1300
Grid size (m)	0.6–6.0			

Table 1. Details of simulation domain size in the four selected areas.

	Low-rise	Middle-rise	Mix-rise	High-rise
Site area (m^2)	123,200	189,280	153,000	95,850
Built area (%)	51	19	14	6
Green area (%)	22	28	41	49
Floor area (%)	106	153	242	254
Average floor	2	5	7	17

Table 2. Land use and building height properties in the four selected areas.

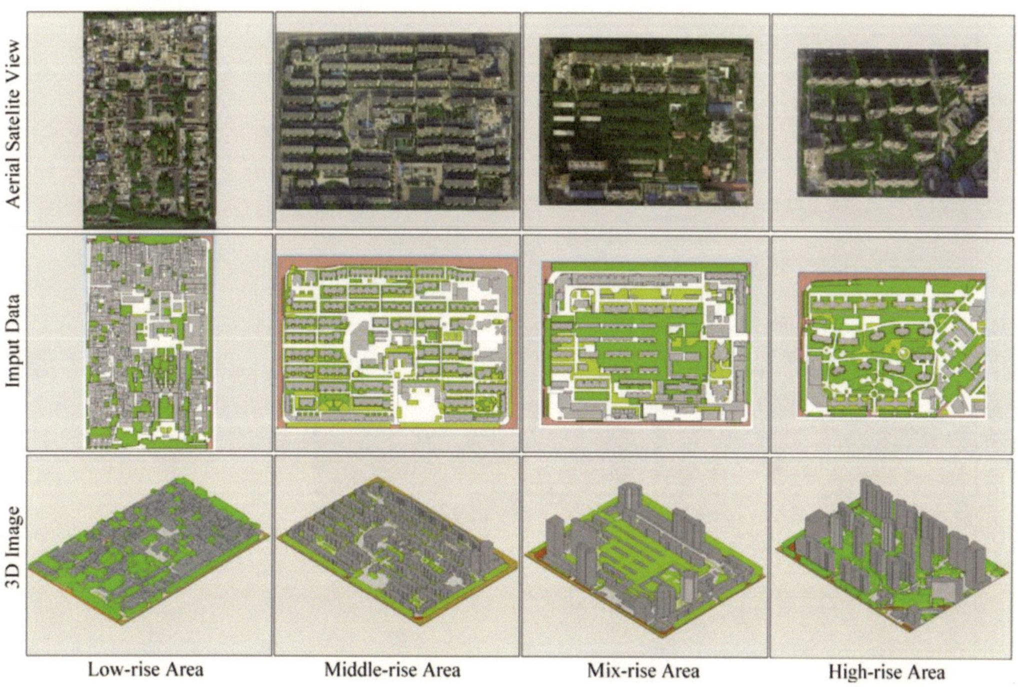

Figure 3. Image selected areas and simulation domain data.

	Satellite image	CAD map	ENVI-met input data
Xiao Zhai	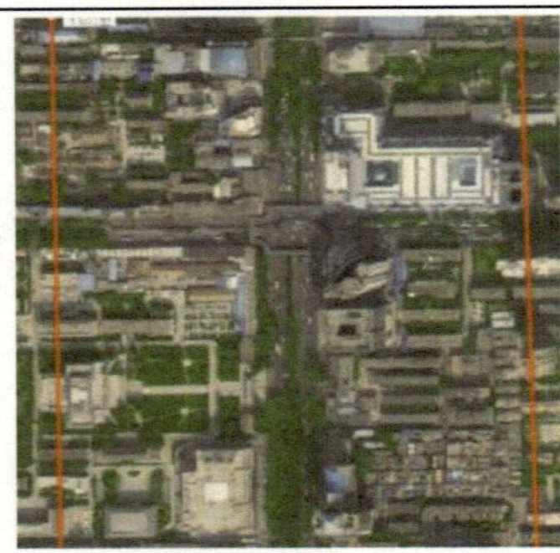	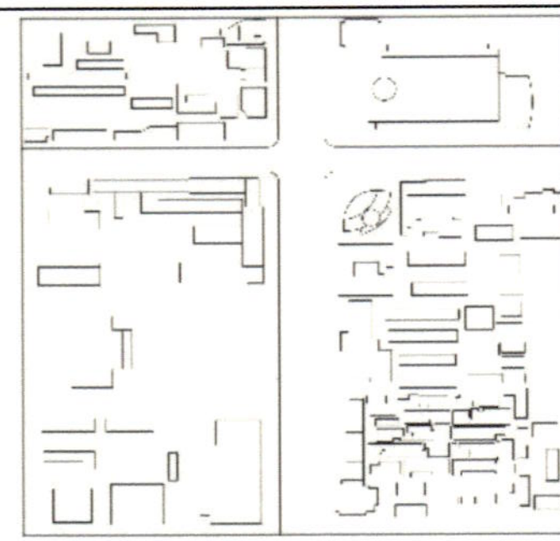	
Shopping district		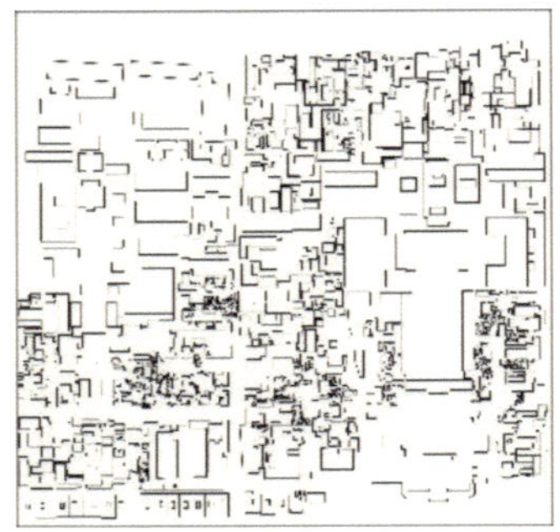	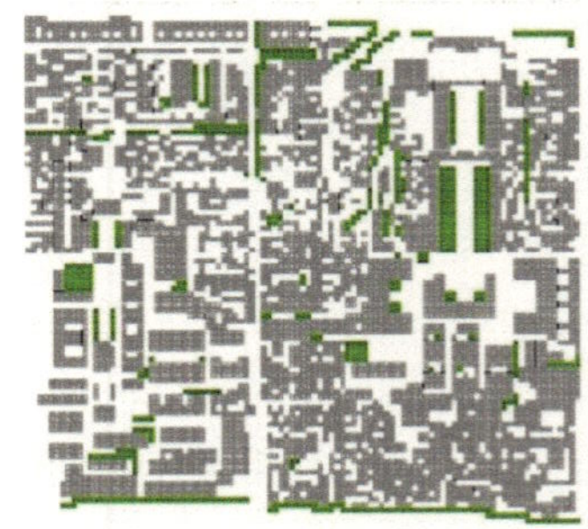

http://dx.doi.org/10.5772/intechopen.79267

	Satellite image	CAD map	ENVI-met input data
Shu Yuan men	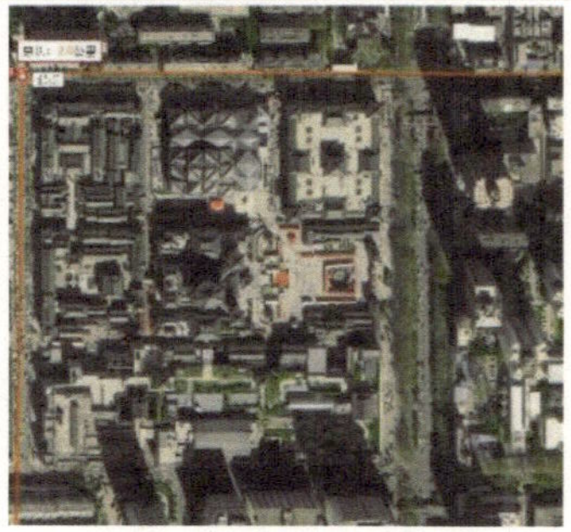	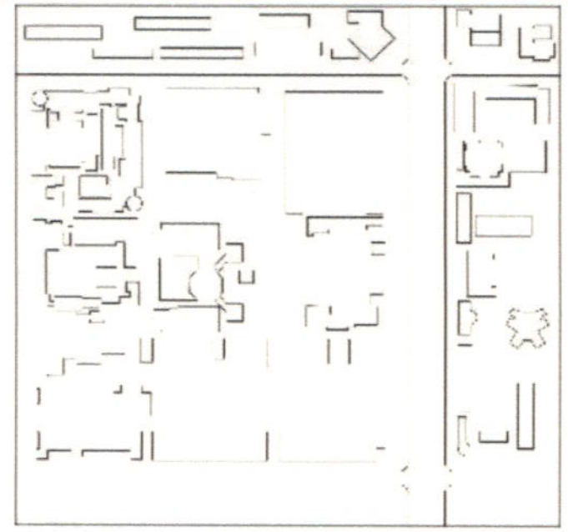	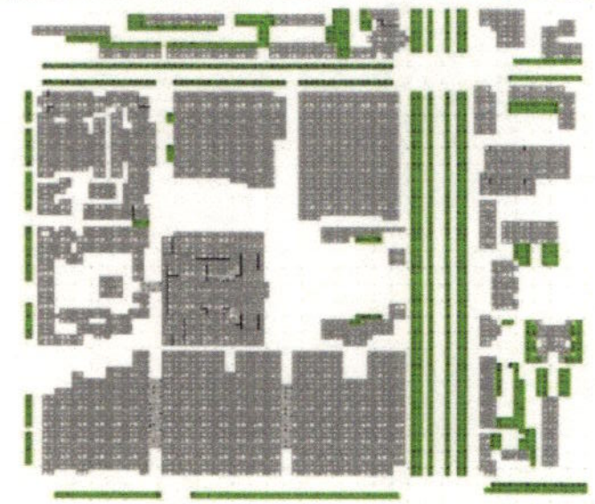

Table 3. Image selected areas and simulation domain data

2.2. Simulation

A commercial CFD code, scSTREAM (Software Cradle Co., 2011), was used to simulate the urban ventilation properties. Detailed distributions of air current and pressure per direction can be visualized. The simulation models were built according to the realities shown in **Figure 3**, and the detailed input data are presented in **Tables 1, 2** and **5**.

The wind environment (Xi'an Weather Station, 2016) on a typical summer day (21st July, 2016) and a typical winter day (21st December, 2015) is selected for analysis. The wind direction and wind speed between 6:00 a.m. and 8:00 p.m. are shown in **Figure 4**. The average wind speed

	Xiao Zhai	Shu Yuan men	Tang West Market Group
Layout type	Layout along the street shopping district	Traditional historic and culture block shopping district	New
Construction time	2002	1991(1906)	2012
Building coverage (%)	32	46	33
Average building height (m)	40.6	5.6	28.3
Green coverage (%)	32.4	24.3	26.6

Table 4. Related parameters of three areas

	Materials	Density [kg/m³]	Specific heat [J/(kg K)]	Thermal conductivity [W/(m K)]
Building	Concrete	1600	1000	0.65
Road	Asphalt	2120	920	0.74
Side walk	Redbrick	1650	840	0.62
Inside surface	Mortar	2000	800	1.3
Green	Soil	1340	1700	0.7

Table 5. Details of ground surface material characteristics for simulation

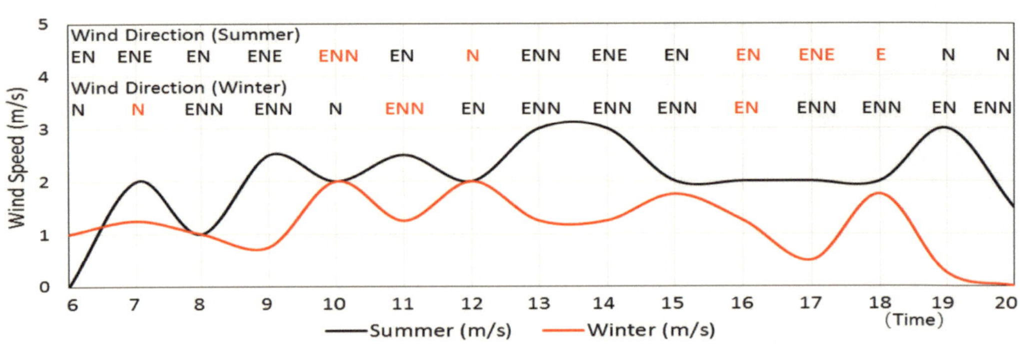

Figure 4. The wind condition in a typical summer day (21st July, 2016) and a typical winter day (21st December, 2015). The main wind directions are shown in red.

on the summer day was 2 m/s, and the wind directions were N, ENN, EN, ENE, and E, the main being EN; the average wind speed on the winter day was 1.25 m/s, and the wind directions were N, ENN, and EN, the main direction being ENN. The average wind speed and all of the wind directions in different seasons served as the initial condition for the simulations in scSTREAM. The main wind directions were here used for detailed results analysis and discussion.

3. Results

3.1. Results of urban air quality and urban ventilation

3.1.1. Air pollution distribution in Xi'an city

The field measurement results of the PM10 concentration distribution in the summer of 2016 are shown in **Figure 5**. The air pollution concentration distribution was not averagely distributed in the city. Comparison of **Figure 4** and the satellite photo in **Figure 2** showed the distribution of air pollution to be partially but directly related to urban density. This is because the urban typology affects urban ventilation and accelerates aggregation, and the air pollution in high-density urban districts is subsequently high.

In order to clarify the mechanism of the air pollution concentration in high-density areas, the low-rise and middle-rise areas that located in highly polluted areas are selected for simulation. This phenomenon will be discussed with the simulation results.

3.1.2. Wind environment simulation

On a summer day, five wind directions (N, ENN, EN, ENE, E) were simulated at a wind speed of 2 m/s. On a winter day, three wind directions (N, ENN, EN) were simulated at a wind speed of 1.25 m/s. **Figures 6** and **7** present the simulation results in the selected four urban areas.

In summer (**Figure 6**), with the effects from trees, the median wind speed in the low-rise area was slower than in the other areas, and the wind speed in the mix-rise area was fastest. With a wind direction of EN, the median wind speed in mix-rise area was 0.35 m/s, which was 0.15 m/s higher than in the low-rise and middle-rise areas, respectively. The median wind speed in high-rise area was slightly slower (0.03–0.06 m/s) than in the low-rise and middle-rise areas, but there were small areas of higher wind speed in the high-rise area, and these reached 2.13 m/s (2–3 times of the max wind speed in the low-rise and middle-rise areas). This could explain the results of high PM10 concentration measured in low-rise and middle-rise areas shown in **Figure 7**.

In winter (**Figure 7**), without the effects from trees, the median wind speed in the mixed-rise area was higher than in the other areas. With a wind direction of ENN, the median wind speed in mixed-rise area was 0.24 m/s higher than in the low-rise area and 0.19 m/s higher than in the middle-rise area. The median wind speed in the high-rise area is 0.09 and 0.28 m/s lower than in the middle-rise and mixed-rise areas. This is to say, the low-rise area (high-built density) and the high-rise area (high urban roughness) are reducing the urban wind speed.

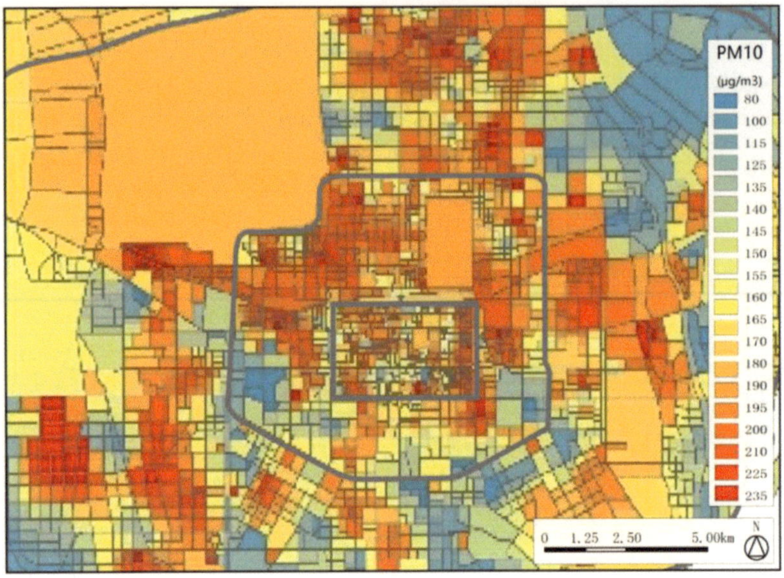

Figure 5. Distribution of PM10 concentration in the central of Xi'an.

In most of the cases, wind direction of EN provides the highest wind speed in allover the area. But, the results of high-rise area in summer are lower than the other directions. This is because of the effects from trees.

3.1.3. *Wind speed distribution*

Figures 8 and **9** show the details of wind speed distribution inside these four areas in the summer and winter. In the low-rise area, because of the high building density and narrow corridors between buildings, the overall wind speed was low, and it also had a pronounced effect on downwind areas. In the high-rise area, some areas of high wind speed were observed in between the high-rise buildings, but the wind speed in the leeside of the big volume buildings was extremely low. The mix-rise area showed the best ventilation properties of the four areas. Therefore, the traditional urban typology with low density and high built coverage creates the low urban ventilation at the human level. High-rise districts with large open space inside the district also reduce the overall wind speed.

In winter, without the effects from the trees, higher ventilation could be observed inside the four districts. Especially in the high-rise district, the wind property inside the community is promoted in the winter.

3.2. Results of urban typology and urban environment in typical shopping areas

ENVI-met was used to calculate the wind speed, the air temperature and the sky view factor (SVF) in all over the area at human height level (1.5 m height from the ground). The results are shown in **Figure 10**. They are all stimulated data at 1400 h.

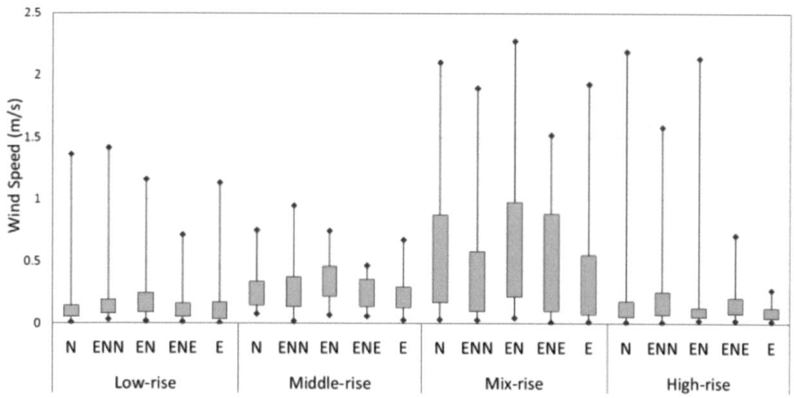

Figure 6. Wind speed in four selected areas in a typical summer day (21st July, 2016).

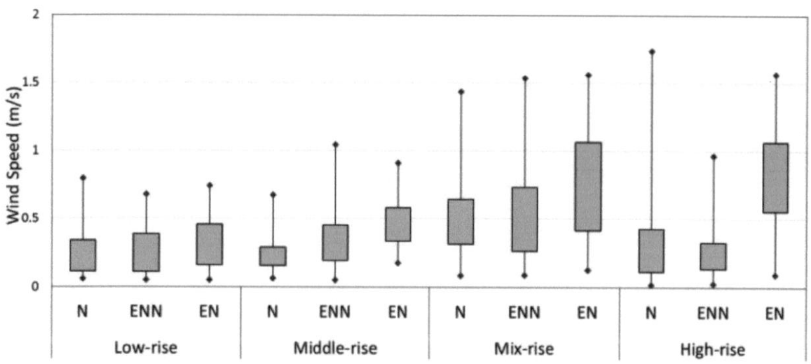

Figure 7. Wind speed in four selected areas in a typical winter day (21st December, 2015).

In **Figure 11**, it is shown the result of the wind speed stimulation of three shopping districts. In a previous paper, Steemers selected six different layouts of building combinations to simulate the wind speed of the regional environment and found when the building parallel to the direction of the wind, ventilation rate is the highest in the street space, but the ventilation rate of building space is poor in the combination of combination and courtyard [13]. As can be seen from the simulation diagram, the average wind speed in Tang West Market Group shopping district (0.77 m/s) is stronger than others (Xiao Zhai is 0.75 m/s and Shu Yuan men is 0.76 m/s). This is because of the lower land cover in Tang West Market Group shopping district which creates more open spaces and wider streets. Meanwhile, the street space in this area is more orderly and vertical. However, Shu Yuan men shopping district's wind speed is stronger than Xiao Zhai, although its site coverage is higher. Because it has less trees in the area and its average building height is lower than Xiao Zhai.

In **Figure 12**, it shows the air temperature of three areas. We can see Xiao Zhai's temperature value is lower than other two. Shu Yuan men's average air temperature is 29°C, and it is 1°C higher than Xiao Zhai's and Tang West Market Group. The proposed reason for this is that more plants are built in Xiao Zhai shopping district, and these trees can effectively reduce

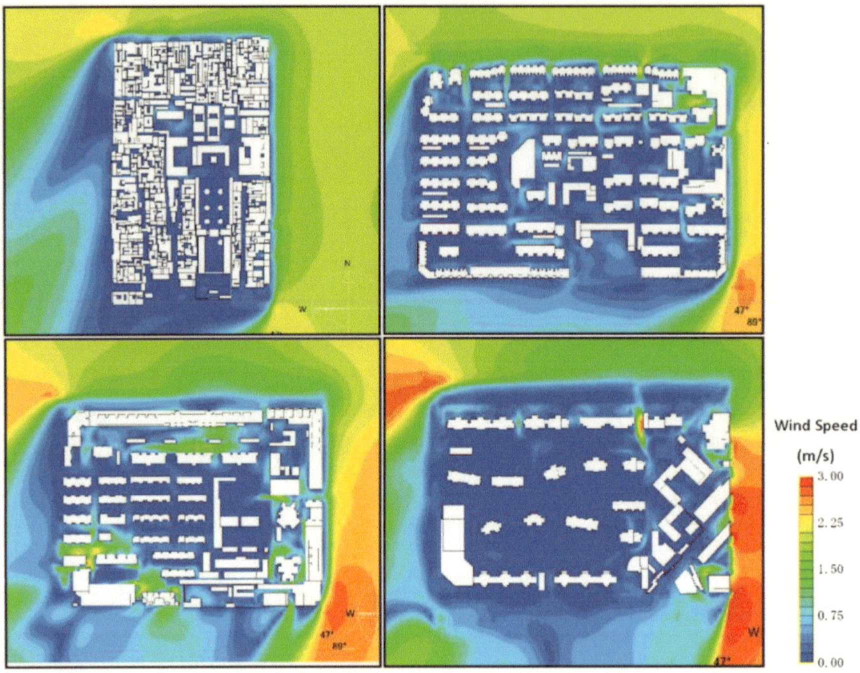

Figure 8. Wind speed distribution in four selected areas at 1.5 m height from the ground in summer (wind speed: 2 m/s; wind direction: EN).

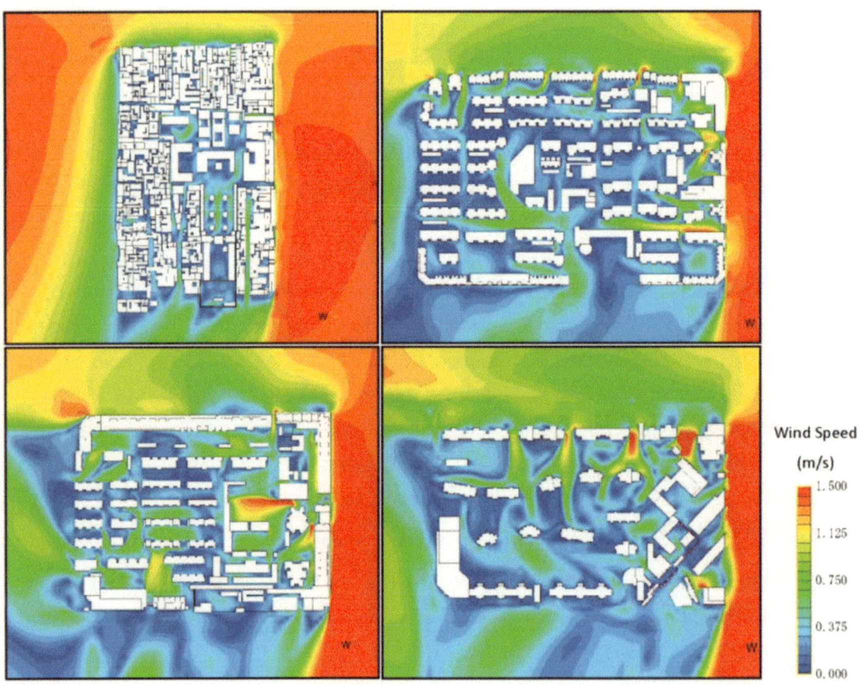

Figure 9. Wind speed distribution in four selected areas at 1.5 m height from the ground in winter (wind speed: 1.25 m/s; wind direction: ENN).

the air temperature in the area. Because the green plants mainly reduce the environment temperature by shadow and evapotranspiration. Through the role of the cover plants, two buildings 'walls and roof surface' temperature can be reduced by 11–25°C [14]. Moreover, the high-temperature area in Shu Yuan men is larger than others because it has less vegetation in the area. This is to say, tree planting in the urban area is providing contribution on wind speed and air temperature reducing, and it can enhance thermal comfort.

In **Figures 10** and **13**, we can see the SVF images of three study areas. View factor is a geometric ratio, which is a part of radiation from the surface A blocked by object B [15]. Unger studied on the 35 city areas for 0.25 km^2 in Hungary Szeged and proved the SVF closely related to the intensity of thermal environment in the microenvironment, which existed a good linear relationship [16]. As the SVF increases, heat intensity decreases [17]. The average of Xiao Zhai shopping district's SVF value is 0.36, but in Shu Yuan men is 0.52 and in Tang West Market Group is 0.45. Therefore, the SVF of Shu Yuan men is the highest of the three.

Figure 10. Sky view factor in the three shopping districts.

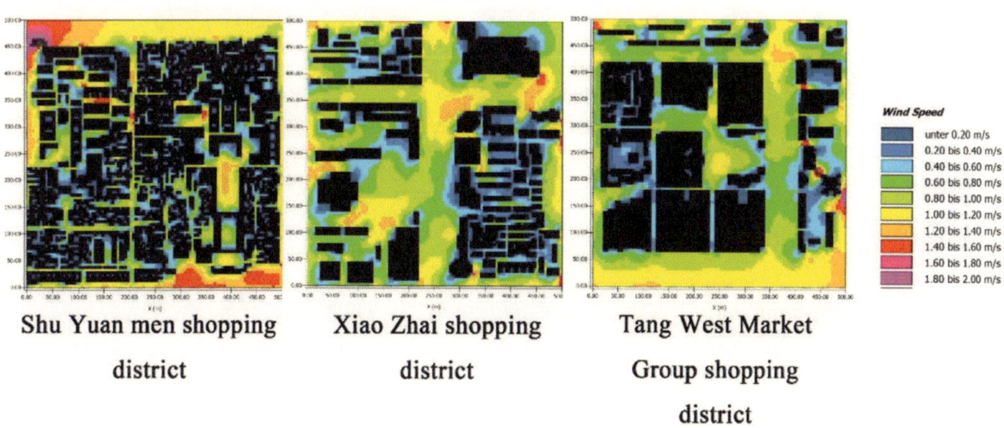

Figure 11. Wind speed distribution in the middle of a summer day (22nd July, 2016) at 1.5 m height.

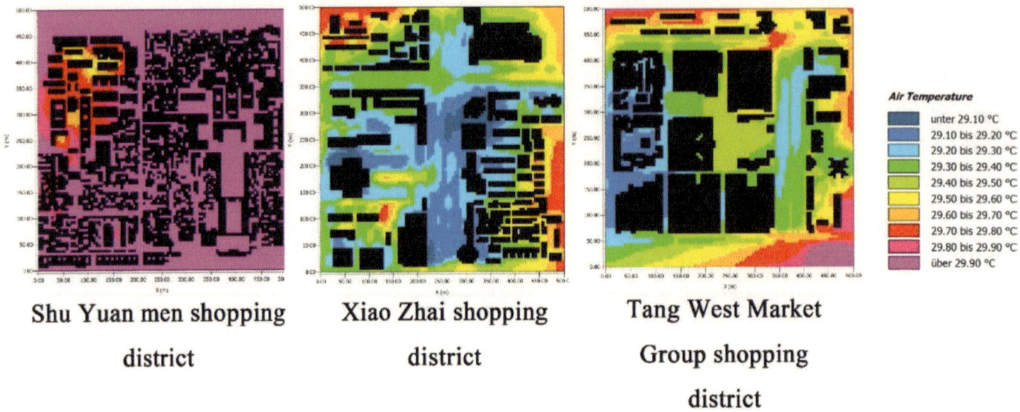

Figure 12. Air temperature distribution in the middle of a summer day (22nd July, 2016) at 1.5 m height.

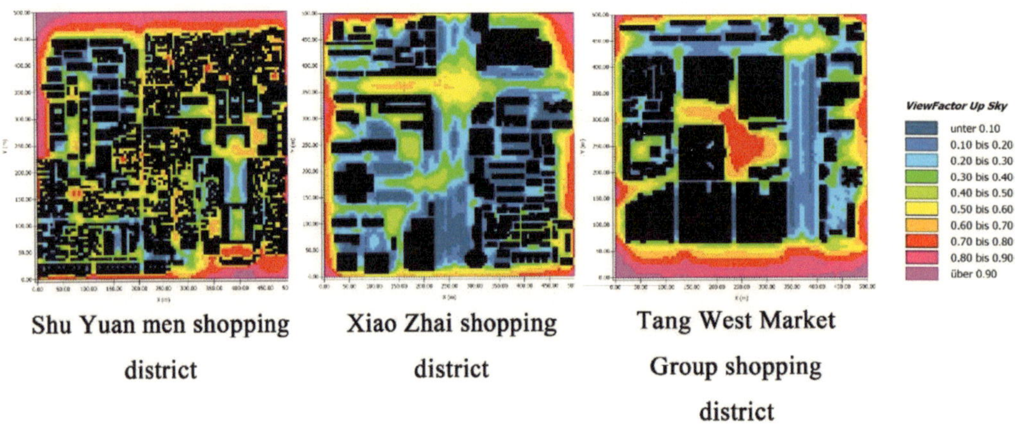

Figure 13. Sky view factors in the middle of a summer day (22nd July, 2016) at 1.5 m height.

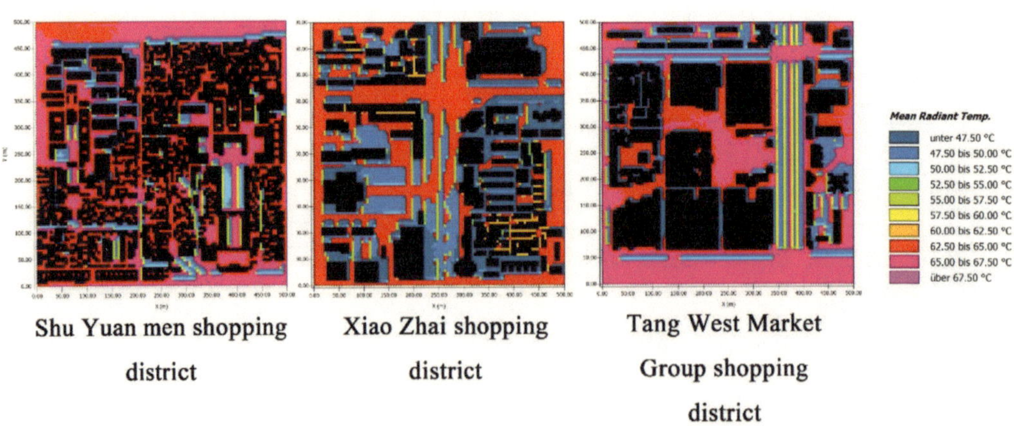

Figure 14. Mean radiant temperature in the middle of a summer day (22nd July, 2016) at 1.5 m height.

This is because the high building density creates deeper urban canopy, and less vegetation in this area model makes more open space. Therefore in Tang West Market Group, we also can see a high numerical concentration in the central area, which is an open square. However, in Xiao Zhai, although there are many open spaces in the area, there are a lot of plants that cover them. Therfore, its SVF value is the lowest of three districts.

In **Figure 14**, these are results of mean radiant temperature in three areas. In these images, we can see that lower values mainly concentrate in the green plants area in the shopping district, and higher values is in the other spaces without vegetation.

4. Discussion

Urban ventilation plays an important role in the urban environment. In summer, urban ventilation contributes to urban heat dissipation and urban heat island mitigation. In winter, high wind speed accelerates the aggregation of air pollution. Results demonstrated that the urban typology affects urban ventilation and urban air quality. However, it is usually difficult to change the urban form over a short period in areas that have already been developed.

Most of China's cities have undergone fast economic growth, urban expansion, and urban redevelopment. This unique situation is the reason of the rapid environmental degradation in Chinese cities, which could also be useful to the other countries. The fast change in form in China's urban areas has provided an opportunity to optimize urban environments in short periods. This requires an urgent establishment of related policies to regularize environmental urban development and redevelopment.

5. Conclusion

This work demonstrated that the wind environment in the low-rise area and the high-rise area are characterized by high building density and the pronounced urban roughness. Wind speed was 0.04–0.09 m/s lower in the high-rise area than in the middle-rise area and 0.04–0.14 m/s lower in the low-rise area than in the middle-rise area. Wind speed is 0.19–0.27 m/s lower in the high-rise area than in the mixed-rise area and 0.21–0.28 m/s lower in the low-rise area than in the mixed-rise area. Overall, the balance between building height and building ratio should be considered in future urban development projects. The information from this work provides information useful to the cultivation of environmental urban policy.

Overall, high-density urban residential and commercial development is providing a big impact on the urban wind environment and urban thermal environment. While this gives hints for UHI mitigation during the day, and it creates physical obstacles for heat release during the nights. A lower SVF reduces the urban radiation absorption from the Sun, but also reduces the outgoing longwave radiation from the urban surfaces. The spread of air pollution is affected by the wind turbulence around high buildings. Future studies should consider more detailes of the layout and volume of high-rise buildings in urban development projects to reduce urban climate change. The results of this study provide clues for related environmental urban development mechanisim in Chinese cities as well as in the other cities in all over the world.

Author details

Wang Yupeng*, Ma Dixuan, Li Man and Zhou Dian

*Address all correspondence to: wang-yupeng@outlook.com

Xi'an Jiaotong University, Xi'an, China

References

[1] Ren GY, Chu ZY, Chen ZH, Ren YY. Implications of temporal change in urban heat island intensity observed at Beijing and Wuhan stations. Geophysical Research Letters. 2007;**34**(5):89-103

[2] Xiao R, Ouyang Z, Zheng H, Li W, Schienke E, Wang X. Spatial pattern of impervious surfaces and their impacts on land surface temperature in Beijing, China. Journal of Environmental Sciences. 2007;**19**(2):250-256

[3] Natural Resources Canada. Climate Change Impacts and Adaptation: A Canadian Perspective. 2004. http://adaptation.nrcan.gc.ca/perspectivee.asp

[4] Hung T, Uchihama D, Ochi S, Yoshifumi Y. Assessment with satellite data of the urban heat island effects in Asian mega cities. International Journal of Applied Earth Observation and Geoinformation. 2006;**8**(1):34-48

[5] Zhan W, Zhang Y, Ma W, Yu Q, Chen L. Estimating influences of urbanizations on meteorology and air quality of a central business district in Shanghai, China. Stochastic Environmental Research and Risk Assessment. 2013;**27**(2):353-365

[6] Bosselmann P, Arens E, Dunker K, et al. Urban form and climate: Case study, Toronto. Journal of the American Planning Association. 1995;**61**(61):226-239

[7] Wang Y, Akbari H. Effect of sky view factor on outdoor temperature and comfort in Montreal. Environmental Engineering Science. 2014;**31**(6):272-287

[8] Wang Y, Berardi U, Akbari H. Comparing the effects of urban heat island mitigation strategies for Toronto, Canada. Energy and Buildings. 2015;**114**:2-19

[9] Wang Y, Akbari H. Analysis of urban heat island phenomenon and mitigation solutions evaluation for Montreal. Sustainable Cities and Society. 2016;**26**:438-446

[10] Xi'an Yearbook Bureau. Xi'an Yearbook. World Publishing Corporation. 2015. p. 10

[11] Lyu Y, Liu K, Niu J, Liu L, Zhao D. Problems and strategies of urbanization from the perspective of city-dominated society—A case study of Shaanxi Province. Economic Geography. 2013;**33**(7):61-68

[12] Lv W. Research on Urban Land-Use Change of China Based on the Urban System. Chongqing: Southwest University; 2013

[13] Ratti C, Raydan D, Steemers K. Building form and environmental performance: archetypes, analysis and an arid climate. Energy and Buildings. 2003;**35**:49-59

[14] Akbari H, Kum DM, Bretz SE, Hanford JW. Peak power and cooling energy savings of shade trees. Energy and Buildings. 1997;**25**:139-148

[15] Unger J. Connection between urban heat island and sky view factor approximated by a software tool on a 3D urban database. Environment and Pollution. 2009;**36**:59-80

[16] Oke TR. Boundary Layer Climates. 2nd revised ed. London: Routledge; 1987

[17] Zheng J. Study on the scheme optimization of urban design based on the wind and heat environment simulation on a block scale—Taking CBD in east of the river of Xiangtan City as an example. Chinese and Overseas Architecture. 2015;**7**:111-115

Chapter 5

Multi-criteria Spatial Decision Support System for Urban Energy Planning: An Interdisciplinary Integrated Methodological Approach

Sara Torabi Moghadam and Patrizia Lombardi

Additional information is available at the end of the chapter

http://dx.doi.org/10.5772/intechopen.80883

Abstract

The present chapter provides an interdisciplinary integrated methodological framework. This framework guides to develop a multi-criteria spatial decision support system (MC-SDSS) to support decision-making processes in urban energy planning (UEP) purposes. The MC-SDSS helps in evaluation and visualization of the results of different UEP scenarios involving the relative stakeholders and decision-makers (DMs) from the early stage of planning. This will help in defining and evaluating energy-saving scenarios taking into account the participation of stakeholders in an interactive way. The meaning of integrating different tools and methods in this framework is due to their complementarity in fulfilling various tasks in the UIEP process. This fact can help to assess, over a short-/long-term period, the mix of measures by analysing meaningful scenarios focused on energy consumptions, environmental impacts and economic and social aspects. The result is the development of a new MC-SDSS, which is an interactive energetic plug-in in GIS environment using CommunityViz. This tool has been applied to a demonstrator case study, related to a medium-sized city of the metropolitan area of Turin. However, the methodology used for delivering the tool can be applied to other contexts due to its flexibility.

Keywords: interactive energy retrofitting scenarios, spatial planning support system: an interdisciplinary integrated methodological approach, geographic information system (GIS)

1. Introduction

Sustainability contests represent a fundamental challenge to traditional urban development practices and concepts. Reducing energy consumption and greenhouse gas emissions from urban infrastructure and building stock towards low-carbon cities requires a supportive planning process. In this regard, the use of appropriate tools and methods in order to address complex interactions of urban integrated energy planning (UIEP) processes is needed [1]. However, there is still not an integrated method to meet the urban integrated energy planning (UIEP) purposes [2]. In the mentioned study, the necessary approaches, which are needed to create the future urban energy consumption paths for scenario analysis, are described. Particularly, the importance of using geographic information system (GIS) for calculating, managing, storing and visualizing data at the urban scale is highlighted.

The chapter discusses in detail the steps design of methodological approach of a new integrated multi-criteria spatial decision support system (MC-SDSS) to evaluate and visualize the results of different UIEP scenarios involving the relative stakeholders and decision-makers (DMs) from the early stage of planning [3]. This will help in defining and evaluating energy-saving scenarios taking into account the participation of stakeholders in an interactive way. The meaning of integrating different tools and methods in this framework is due to their complementarity in fulfilling various tasks in the UIEP process.

The proposed methodology framework is explained in Section 2. Afterwards, Section 3 illustrates the first results of the methodology application. Finally, some concluding and limitation remarks of this study are given in Section 4.

2. Methodology: an interdisciplinary integrated approach

From the literature basing on a compilation of fragmented definitions, the section puts forward a synthetic description of key terminologies used, in order to facilitate and improve the debates on this emerging field [4]. UIEP is defined by [1], as a model-based energy planning process. This is divided into the following four main phases: Phase I, preparation and preliminary analysis; Phase II, detailed urban buildings energy modelling; Phase III, prioritization and decisional process and Phase IV, implementation and monitoring (**Figure 1**).

To address a complex issue of UIEP, which consists of many different planning phases involving multi-sectors and objectives, there is a need for an interdisciplinary approach [5]. In fact, the planning processes in urban energy problems may be not specifically innovative approach; however, its management by means of integrated, cross-sector, multi-criteria and multi-actor approaches is absolutely a novel approach to be resolved [4]. In this vein, many cities should struggle to develop innovative methods to successfully reinforce the collaboration among different research disciplines dealing with energy issues [6].

On one hand, considering the existing research gaps and methodological directions, this study follows an interdisciplinary path. Both technical (e.g. energy modeling) and societal (e.g. an active engagement of relevant actors and interest groups) elements help to perform a proper UIEP, especially from the stakeholders' perspective [7]. On the other hand, this study

http://dx.doi.org/10.5772/intechopen.80883

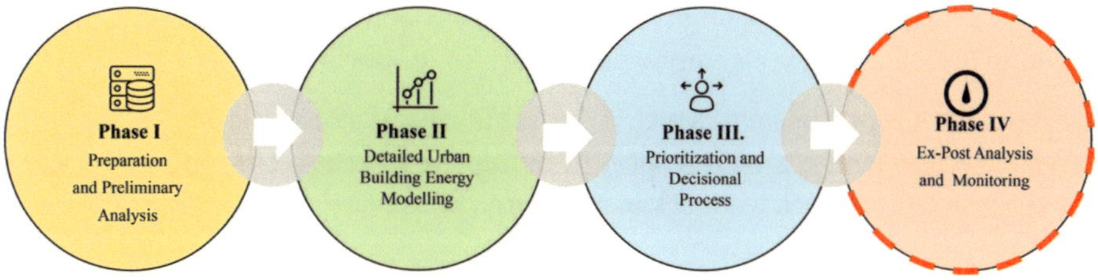

Figure 1. Urban integrated energy planning (UIEP) phases, adopted from [1].

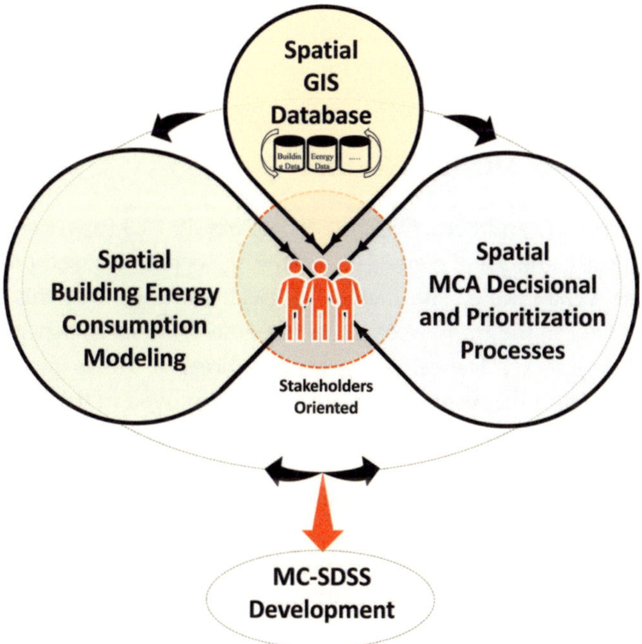

Figure 2. Schematic overview of the three main components of this research.

is fundamentally based on 'multi-methodology integration' defined by [1], in which parts of different methodologies are combined (e.g. statistical, engineering, focus groups, etc.).

In structuring the UIEP, it is important to select different appropriate approaches and to choose them considering the decision context and the type of planning project. Furthermore, it is crucial to analyse how it is possible to implement the interaction among the different stakeholders. As a result, the developed MC-SDSS for UIEP in the built environment uses techniques at the crossroads of three domains (see **Figure 2**):

- Spatial database, which constitutes the GIS platform including all the relative information and data and enables the use of analytical process and outcomes such as the maps, graphs and tables

- Spatial building energy modeling, which develops a bottom-up modeling to evaluate the current and future energy consumption at the city scale concluding a sufficient level of detail
- Spatial decision support system, which is the fact that the decision-makers (DMs) can express and exert their preferences with respect to multiple evaluation criteria and/or alternatives and, consequently, get back feedback in a real time to increase the DMs trust in the outcomes

The integration and combination of this technical know-how allow providing maps of energy, economic, environmental, social and technical indicators resulting from the evaluation of energy-saving scenarios. This provides a supportive tool for the urban actors in the participatory planning processes allowing several stakeholders with different backgrounds and interests to gather and discuss the issues of several urban and regional energy-saving scenarios [8]. In the following section, the integration of theoretical proposed framework and how it is supposed to be applied to the study practice are shown.

2.1. Research framework of MC-SDSS for UIEP

A new MC-SDSS, which is an interactive plug-in of ArcGIS 10.3 (www.arcgis.com) environment helps dynamically analyse the energy retrofitting scenarios based on the stakeholders' preferences over an urban scale. The methodological framework of this study consists of several phases involved in the framework of an integrated urban energy planning according to Mirakyan and De Guio [1]. Hence, it is helpful to break it down into the main elements that frame it to understand the research process steps employed in this study. To this end, in **Figure 3** a schematic flowchart of the methodological approaches is shown.

2.1.1. Phase I: preparation and preliminary analysis

Accordingly, the fieldwork should be started from the quantitative data collection to characterize the building stock and to create a supportive geodatabase. This phase (Phase I) is entitled 'preparation and preliminary analysis'. Phase I is the foundation of all processes and modeling approaches in the next Phases, II and III. Of course, the GIS database can be always updated, and more data can be joined into the framework. In this step, the information characterizes by georeferenced and non-georeferenced data. Therefore, the georeferencing procedure should be performed for those non-georeferenced ones in order to create a strong geospatial database. All the collected data have been then overlapped and integrated into the GIS platform. In this regard, each building polygon has been associated with its available and necessary information. The goal of this phase is to create a 2D-GIS-database platform for the city including the several factors, which may influence the building energy issues. The use of GIS was crucial since it offers the opportunity to characterize the building stocks and to visualize the spatial distribution of a large number of data through its location-based feature and its multiple layers representation.

2.1.2. Phase II: detailed urban buildings energy modeling

Consequently, Phase II intends to perform to model the energy consumption of building stock in a detailed way. First, a bottom-up statistical model has been developed to estimate

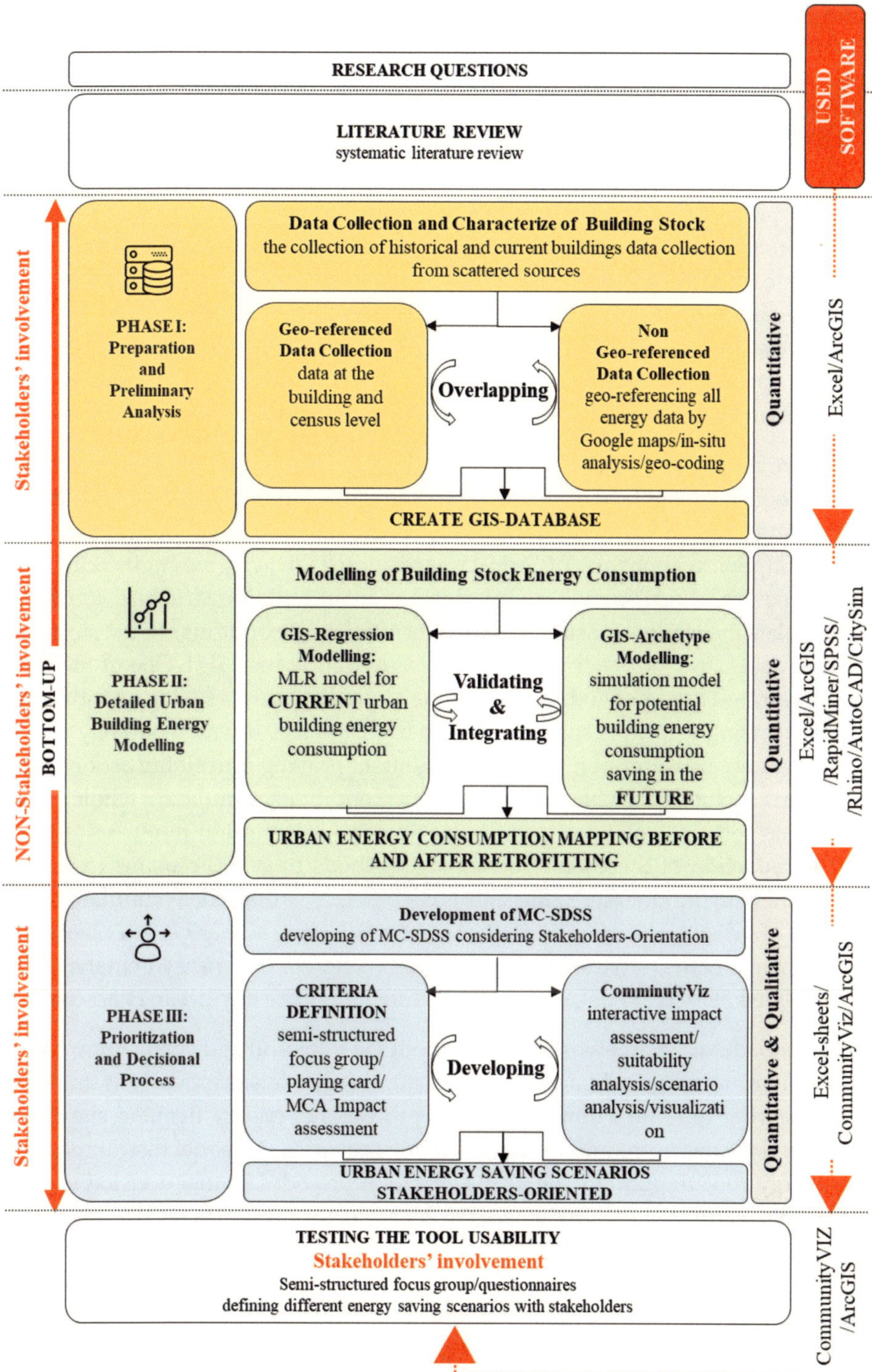

Figure 3. A schematic overview of the methodological approach.

the heating energy consumption for the built environment space heating at the city level. This model is based on the integration of statistical analysis with 2D-GIS to map the current energy consumption of the city [9]. The novelty of the proposed statistical model lies on its simplicity and applicability and the high level of their robustness in the literature. However, these statistical methods rely strongly upon monitored real data. It should be noted that fortunately, the author succeeded to collect a sample of information of energy billings as a data source for modeling purpose and for analysing the link between energy consumption and a wide range of different variables. Moreover, the statistical models are also able to take into account socio-economic effects in the equations [10]. They perform reliable consumption information about the present condition of buildings and for the calibration process of engineering-based models. However, due to the strong dependency of statistical methods on available historical consumption data, these methods are limited to predict the impact of innovative technology options and energy-saving potential after applying renewal solutions.

In counterpart, engineering methods are very detailed models based on traditional thermodynamic relationships and heat transfer calculations [10]. Although the historical data can be used for making a comparison against measured consumption data, these methods can assess energy consumption without any historical information. However, the engineering modeling approaches require a high quantity of information about building structure and parametric input to calculate the energy consumption of a set of reference buildings of the stock based on a numerical model. In this, 3D city models can significantly help [11]. One of the main benefits of engineering-based methods is their ability to predict energy-saving quantity for buildings after renovating solutions application [12]. In this phase, the methodology proposes to simulate the energy consumption of urban areas after applying retrofitting actions. Although the engineering methods are able to predict future conditions, simulating whole cities using energy demand software can be very extensive in terms of computer resources and data collection. The reduction of these time-consuming methods thus still remains to be resolved. Therefore, a new methodology, using city archetypes, is proposed to simulate the energy consumption of urban areas including urban energy planning scenarios. The objective of this part is to present an innovative solution for the simulation of the energy demand of cities by using a simplified 3D-GIS model, designed as a function of the city urban characteristics.

In fact, the methodology framework combines both the statistical and engineering approaches to obtain a more robust prediction of the urban energy consumption. The framework is performed in order to reduce time-consuming processes of energy demand simulation and assessment and for designing urban energy-saving scenarios. A spatial distribution of urban building energy consumption in 2D/3D visualization provides spatial decision support system (SDSS) tool in order to identify where the energy consumption is mostly concentrated to make the better decisions.

2.1.3. Phase III: prioritization and decisional process

Phase III of the study follows 'a mixed methodology' that combines qualitative and quantitative approaches [7]. Qualitative research refers to semi-structured focus groups formed in which the qualitative data such as stakeholders' opinion are collected through discussions and questionnaires. Particularly, the use of focus groups by stakeholders in this study has the following implication: it reflects the 'mixed methodology' choice for Phase III with the use of qualitative (semi-structured focus groups, questionnaires, playing card) and quantitative (building stock energy data, costs, etc.) data collection and analysis methods.

Regarding the definition of evaluation criteria, several methods exist in the literature [13]. The present study proposes a participative approach in order to define the evaluation criteria through the multi-stakeholders workshop including semi-structured focus group organized involving relevant stakeholders [14]. The definition of evaluation criteria side by side with the real local stakeholders leads to have trustable results that grantee the robustness of planning process. A vast number of available MCDA approaches make it necessary to carefully select the most appropriate method for each specific decision context. In this framework, the 'playing cards' is chosen [15] due to some reasons. First, it is a simple and intuitive method and easy to be understood, even by non-experts in the field of decision processes [16]. Second, they can help DMs in managing values that cannot be quantified without difficulty, involving qualitative judgments. Finally, the technical parameters involved in the playing card methodology can be interpreted easily, allowing a simplification of the problem. Lombardi et al. [14] describe the main the procedure of 'playing cards' method and its results.

Subsequently, each of the selected criteria from the workshop has been analysed and assessed to be implemented in a new MC-SDSS tool (see Section 3). Two main instruments, Interactive Impact Assessment (IIA) and Suitability Analysis (SA), are modeled and adapted in order to develop a new MC-SDSS. Several dynamic attributes and indicators were modeled and coded using CommunityViz as a planning support system (PSS) tool [17]. This PSS tool is selected as a base for further modeling processes due to its several strengths. It helps in analysing and understanding the potential alternatives and their impacts through visual investigation and scenario analysis. Moreover, this tool is interactive and provides dynamic feedbacks on changing the assumptions and viewing the influences of changes on the future scenarios on the fly. Furthermore, it engages stakeholders in participative and collaborative decision-making processes through visualization in real-time approach. All the above strengths lead to stronger consensus and better decisions in resolving complex problems. The detailed methodological procedure developed for supporting this phase of research is under progress of publication.

This methodological approach provides a significant innovative progress in the research field that is developing an interactive plug-in tool for UIEP in the GIS environment. In this regard, finally, the second workshop is organized to test the usability and validate the tool from the real stakeholder point of view. For evaluation purposes, this workshop is included in two semi-structured focus groups. This step attempts to understand the weaknesses and strengths of the mentioned framework. In this workshop, the questionnaires also were designed for analysing the stakeholders' feedbacks about the developed tool. Within the use of this GIS extension, public administrative users, such as urban energy planners, policymakers and built environment stakeholders, can plan, design and manage low-carbon cities. This plug-in will provide the stakeholders with the ability to visualize interactively and explore a range of possible future-saving scenarios.

3. Methodology application

The present section illustrates the interface of the developed MC-SDSS tool based on the methodology framework explained in the previous section (Section 2). As mentioned before, this tool is an interactive plug-in in GIS environment, which has been adapted from an existing urban planning tool called CommunityViz. The developed MC-SDSS tool supports the stakeholders in urban energy planning through participatory and collaborative processes. It helps make better decisions by expressing the stakeholders' preferences and their conflicting objectives.

3.1. An interactive MC-SDSS tool

All the phases were integrated in order to create a new MC-SDSS tool. This tool uses CommunityViz, which is an ArcView modular GIS-based decision support system developed by the Orton Family Foundation (http://www.communityviz.com). The above-said tool is able to integrate different types of data such as scripts, numbers, 2D maps, 3D visualization and raster in a real-time and multidimensional environment [17]. CommunityViz encompasses two main components as extensions to ArcGIS: (i) Scenario 360 to map and analyze and (ii) Scenario 3D to visualize. Conceptually, Scenario 360 can be described as a spatial spreadsheet allowing for calculations on spatially related data and formulas that call standard GIS functions [18]. Since each formula, assumption and dependency is viewable and editable, there is not any 'black box' element to a model defined in Scenario 360 [18].

CommunityViz Scenario 360 adds interactive analysis tools and a decision-making framework to the ArcGIS platform with which stakeholders can understand the planning processes easily.

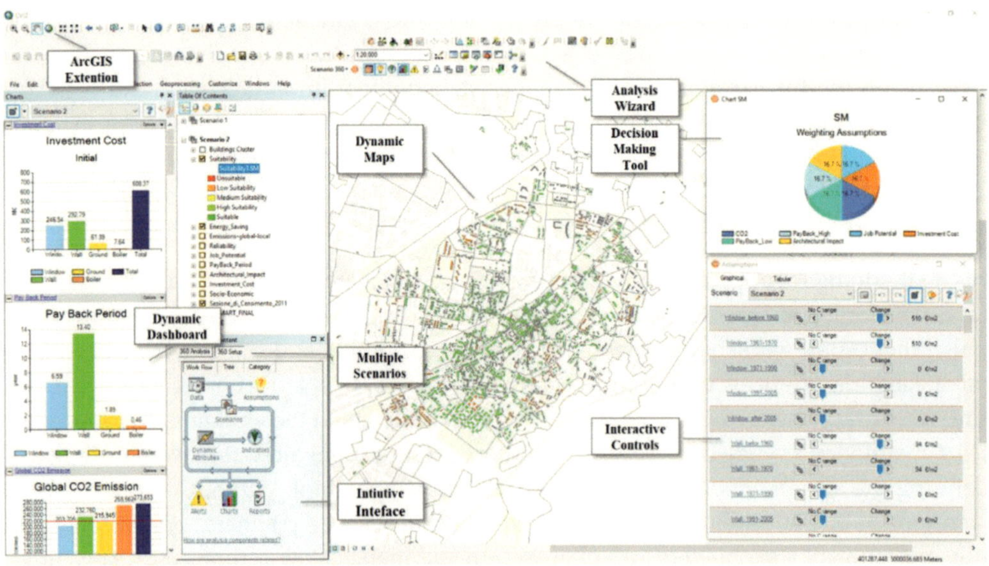

Figure 4. CommunityViz interface; the case study of Settimo Torinese.

Stakeholders can define different decision assumptions and visualize on the fly how the changes may affect environmentally, economically, technically and socially the future scenarios. This dynamic process helps urban actors to negotiate in order to make better decisions [17]. Moreover, it helps facilitate an understanding of the complex problems such as UIEP [19]. Within this tool, many presentation features are available to assist in sharing information with the users including maps, alerts and charts. In these views, stakeholders can ask 'what-if' questions and visualize 'if-then' scenarios in a real time and discuss it very quickly and effectively [20].

CommunityViz Scenario 360 is selected for this study due to its several strengths. It helps analyze and understand the potential alternatives and their impacts through visual investigation and scenario analysis. Moreover, it creates a real-time experiment with different scenarios, changing the assumptions quickly and viewing influences on changes. Furthermore, it engages stakeholders in participative and collaborative decision-making processes through visualization and interactive media [21]. All aforementioned strengths lead to stronger consensus and better decisions in resolving complex problems.

Figure 4 shows the interface of Scenario 360 modeled for the case study of Settimo Torinese. Particularly, the tool consists in building dynamic attributes, which are changeable based upon:

- Data: dynamic data layers create new or add existing layers to the Scenario 360 analysis geodatabase. An important feature is that it provides a dynamic data about features on a map that can be performed by formulas. Therefore, when one aspect changes, the software recalculates the entire analysis.
- Assumption: slider bars or tables let change assumptions during analysis. Using the assumptions, the stakeholders can express their preferences and decisions. When an assumption is changed, all associated formulas with that assumption are automatically recalculated within the scenario.
- Indicators: formula-driven analysis results that are updated automatically while the analysis is performed. Indicators can show the outcome of one or several dynamic attributes.

The stakeholders can experiment their preferences and decisions altering the slide bars. Consequently, they can visualize and analyse the impact of their decisions over different energy retrofitting scenarios. The impact of different scenarios is then visible through different charts, maps and indicators. The tool provides the ability of comparison among different scenarios and indicators.

4. Conclusion

This chapter summarizes the overall conclusion and relative limitation for each phase of planning. In particular, this work creates a link between energetical, economical, societal, technical and environmental performances of retrofitting interventions. The research boundaries were delineated by focusing on existing residential building stock since they characterize the context of most European cities. The relative available data of these buildings were first collected and georeferenced from various sources. Based on the created geospatial database, the building energy consumption patterns were statistically modeled to map the current energy patterns over the entire city. Afterwards, the archetype model of the city was created in order to speed up and ease the future energy-saving simulations by applying the retrofitting solutions. The geospatial database was used as the object of multi-criteria analysis assessments. Finally, an interactive MC-SDSS was created to support the DMs in defining energy-saving scenarios in real time.

4.1. Phases I and II

As illustrated in **Figure 3**, to model the energy consumption over the entire city, a large number of historical data are needed. The most challenging issue was related to collecting and integrating the built environment data and information since the data are significantly scattered among several entities at the local level, and there is a lack of interoperability among the data sources. Actually, this section reports that one of the main barriers to developing a robust and detailed analysis is correlated with the data collection procedure. Especially in Italy, information about building stock and their energy performances are derived from different regional and local authorities, and they are not often homogeneous. Therefore, in order to set up an effective energy planning at the local scale, it is crucial to improve the quality of data availability and management. Data availability of buildings' energy consumption will hopefully improve in the future thanks to smart metering and real-time data monitoring following recent open data policy. To this end, a supportive GIS database where all the scattered information and data were georeferenced is first created.

Referring to the energy consumption modeling at the urban scale, this framework proposed a geospatial statistical modeling. Generally, statistical methods estimate the energy consumption based on a historical data. The model succeeded to estimate the energy consumption of most existing buildings, where the monitored data was not available. However, due to the strong dependency of statistical models on existing available data, these methods are not able to predict the impact of the future refurbishment solutions. Therefore, there was a need to simulate the future city energy performances. However, the simulation of the whole city may be extremely time-consuming.

Therefore, Phase II within the research framework in **Figure 3** proposed a novel engineering methodology to accelerate the urban area energy consumption simulations, including urban planning renovation scenarios. The energy demand of cities, as well as the microclimatic conditions, was calculated by using a simplified archetype 3D model designed as a function of the city urban characteristics. By the proposed archetype modeling approach, this method shows that the number of buildings to be simulated can be drastically reduced with no particular influence on the accuracy of the results. On one hand, the main advantage of an engineering-based method is the capability of predicting energy savings for buildings after the application of renovation measures. On the other hand, these methods are very detailed models based on thermodynamic relationships and heat transfer calculations. As a general remark, the historical data can be used for the comparison against measured consumption data.

4.1.1. Limitations

This study suggests first a spatial data collection and then an integrated procedure of urban energy modeling approaches based on the data collected (i.e. statistical and engineering). This faces several barriers including:

Regarding the data collection

- The energy consumption data is not usually open source; thus, a huge effort was needed to collect the data from different entities and to ask the collaboration from local stakeholders.
- The georeferencing procedure of data could be also a challenging issue. In many cases, the necessary information related to the buildings are associated with the building number (as points) rather than the building polygon. The tricky issue is that these points are sometimes situated between two or three buildings having the same distance. Thus, it is not easy to understand that the data belongs precisely to which building.

Regarding the statistical modeling approach

- A vast amount of historical available data is needed. For many regions, it is almost impossible to have a monitored data in terms of energy performances.
- The intrinsic limitation of statistical methods concerns the microclimate effects, which were not taken into account in the present work. In fact, a microclimate model that would give a single value for the whole city for air temperature would not significantly improve the results of the current model presented in this work.

Regarding engineering modeling approach

- The need for high-level detailed thermo-physics data of the buildings in the city.
- Setting up the simulations can be a tedious task requiring a lot of time and expertise.
- The simulations themselves are very time-consuming, and they require high-performing processors in order to perform the entire city.

4.2. Phase III

As illustrated in **Figure 3**, a MC-SDSS has been developed to support the stakeholders with different backgrounds and preferences. The tool is an interactive plug-in in ArcGIS environment. MC-SDSS is able to help participants in a user-friendly way to define energy refurbishment scenarios. Moreover, the tool gives an opportunity to generate the suitability maps, with which the stakeholders can analyze the grade of the suitability of their decisions. The development of MC-SDSS is based on an existing tool, named CommunityViz. Originally, CommunityViz is a software used to support urban planning purposes. Within this research, CommunityViz was adapted and modeled to support UIEP. Two main integrated instruments, Interactive Impact

Assessment and Suitability Analysis, were modeled. The main difficulty was to adapt the tool to energy urban planning, considering many complex aspects of this issue. Modeling of all retrofit dynamic attributes and the type of connection between all the attributes was another difficulty of this part. The modeling design process is quite complex. The model should chain all the data, attributes and indicators. This means that once the stakeholders change one parameter, others will change automatically in their proposed scenario. The participants are able to rapidly experiment different energy renovation scenarios and change the assumption. This creates an effective interaction between the stakeholders. They can visualize very complex problem of energy-saving scenarios simply by different dynamic colorful maps, charts and indicators.

4.2.1. Limitations

This study suggests the development of a new MC-SDSS, which can define dynamic retrofitting scenarios side by side with stakeholders. This faces several barriers including:

- The need for the tool that to be open source
- The difficulty of the workshops to be time-consuming involving real stakeholders
- The difficulty of the inclusion of conflicting point of views and then aggregation of stakeholders' preferences in a participative decision-making context

The proposed framework will help urban actors to develop energy planning projects, guiding them in the choice among a considerable number of existing planning approaches. The main advantages of the developed MC-SDSS in the field of urban energy planning can be summarized as follows:

- To allow the participative processes
- To give a visualization opportunity for the decision process in specific areas
- To consider multiple criteria (e.g. economic, environmental, technical and, particularly, social aspects)
- To manage and store a very large amount of georeferenced data and to illustrate results requested by users according to different spatial forms (e.g. maps, graphs)
- To show the distribution of buildings' geometrical characterization and buildings' energy consumption

Acknowledgements

The authors of this study wish to acknowledge the National Cluster Smart City and Communities project, named 'EEB-Zero Energy Buildings in Smart Urban Districts' funded by National Operational Program for Research and Competitiveness 2007–2013 (PON R and C), CTN01_00034_594053.

Conflict of interest

The authors declare no conflicts of interest.

Other declarations

The present chapter is emerged from the 3 years research of PhD thesis of Sara Torabi Moghadam under the supervision of the Professor Patrizia Lombardi and co-supervision of the Professor Guglielmina Mutani in order to illustrate the methodological framework used to create an interactive MC-SDSS.

Author details

Sara Torabi Moghadam* and Patrizia Lombardi

*Address all correspondence to: sara.torabi@polito.it

Interuniversity Department of Regional and Urban Studies and Planning (DIST), Polytechnic University of Turin, Turin, Italy

References

[1] Mirakyan A, De Guio R. Integrated energy planning in cities and territories: A review of methods and tools. Renewable and Sustainable Energy Reviews. 2013;**22**:289-297. DOI: 10.1016/j.rser.2013.01.033

[2] Torabi Moghadam S, Delmastro C, Corgnati SP, Lombardi P. Urban energy planning procedure for sustainable development in the built environment: A review of available spatial approaches. Journal of Cleaner Production. 2017;**165**:811-827. DOI: 10.1016/j.jclepro.2017.07.142

[3] Moghadam ST. A new integrated multi-criteria spatial decision support system for urban energy planning in the built environment.PhD Theis, Doctoral Program in Urban and Regional Development (30th Cycle) Politecnico di Torino. 2018. p. 277

[4] Cajot S, Mirakyan A, Koch A, Maréchal F. Multicriteria decisions in urban energy system planning: A review. Frontiers in Energy Research. 2017;**5**:1-25. Article 10. DOI: 10.3389/fenrg.2017.00010

[5] Brömmelstroet MT, Silva C, Bertolini L. Assessing Usability of Accessibility Instruments. 2014. Available from: http://dare.uva.nl/search?metis.record.id=430482

[6] Zanon B, Verones S. Land use policy climate change, urban energy and planning practices: Italian experiences of innovation in land management tools. Land Use Policy. 2013;**32**:343-355. DOI: 10.1016/j.landusepol.2012.11.009

[7] Dantsiou D. (De)constructing and transforming workplace practices: Feedback as an intervention. PhD Theis, University of Cambridge Department of Architecture. 2017. p. 251

[8] Girardin L, Marechal F, Dubuis M, Calame-Darbellay N, Favrat D. EnerGis: A geographical information based system for the evaluation of integrated energy conversion systems in urban areas. Energy. 2010;**35**:830-840. DOI: 10.1016/j.energy.2009.08.018

[9] Torabi Moghadam S, Toniolo J, Mutani G, Lombardi P. A GIS-statistical approach for assessing built environment energy use at urban scale. Sustainable Cities and Society. 2018;**37**:70-84. DOI: 10.1016/j.scs.2017.10.002

[10] Mastrucci A, Baume O, Stazi F, Salvucci S, Leopold U. A GIS-based approach to estimate energy savings and indoor thermal comfort for urban housing stock retrofitting. In: Fifth German-Austrian IBPSA Conference (BauSIM 2014), Aachen, Germany, Sept. 2014. p. 22-24

[11] Nouvel R, Mastrucci A, Leopold U, Baume O, Coors V, Eicker U. Combining GIS-based statistical and engineering urban heat consumption models: Towards a new framework for multi-scale policy support. Energy and Buildings. 2015;**107**:204-212. DOI: 10.1016/j.enbuild.2015.08.021

[12] Mangold M, Österbring M, Wallbaum H. Handling data uncertainties when using Swedish energy performance certificate data to describe energy usage in the building stock. Energy and Buildings. 2015;**102**:328-336. DOI: 10.1016/j.enbuild.2015.05.045

[13] Wang JJ, Zhang CF, Jing YY, Zheng GZ. Using the fuzzy multi-criteria model to select the optimal cool storage system for air conditioning. Energy and Buildings. 2008;**40**:2059-2066. DOI: 10.1016/j.enbuild.2008.05.011

[14] Lombardi P, Abastante F, Moghadam ST. Multicriteria spatial decision support systems for future urban energy retrofitting scenarios. sustainability, MDPI. 2017:1-13. DOI: 10.3390/su9071252

[15] Simos J. Evaluer L'impact sur L'environnement: Une Approche Originale par L'analyse Multicritère et la Nègociation. Lausanne: Presses Polytechniques et Universitaires Romandes; 1990

[16] Figueira J, Greco S, Ehrgott M. Multiple Criteria Decision Analysis: State of the Art Surveys. Springer; 2005. DOI: 10.1007/b100605

[17] Kwartler M, Bernard RN. CommunityViz: An Integrated Planning Support System, Planning Support Systems Integrating Geographic Information Systems Models and Visualization Tools. ESRI, Inc. 2001

[18] Janes GM, Kwartler M. Communities in control: How communities develop their own models using CommunityViz. In: Brail RK, editor. Planning Support Systems for Cities and Regions. Lincoln Institute of Land Policy. 2007. pp. 167-183

[19] Wang JJ, Jing YY, Zhang CF, Zhao JH. Review on multi-criteria decision analysis aid in sustainable energy decision-making. Renewable and Sustainable Energy Reviews. 2009;**13**:2263-2278. DOI: 10.1016/j.rser.2009.06.021

[20] Pelzer P, Arciniegas G, Geertman S, Lenferink S. Planning support systems and task-technology fit: A comparative case study. Applied Spatial Analysis and Policy. 2015; **8**:155-175. DOI: 10.1007/s12061-015-9135-5

[21] Eikelboom T, Janssen R. Collaborative use of geodesign tools to support decision-making on adaptation to climate change. Mitigation and Adaptation Strategies for Global Change. 2017;**22**:247-266. DOI: 10.1007/s11027-015-9633-4

Section 3

The Role of Transport in a Sustainable City

Chapter 6

Electric Two-Wheelers, Sustainable Mobility and the City

Stefan Bakker

Additional information is available at the end of the chapter

http://dx.doi.org/10.5772/intechopen.81460

Abstract

Congestion, lack of accessibility, social equity, air pollution, as well as rising CO_2 emissions are some of the key issues urban transport policymakers face. Motorised two-wheelers and their accessibility benefits are often ignored in this discourse or seen as undesirable for reasons of pollution, noise, road safety and driving behaviour. Cycling, on the other hand, is viewed positively, yet faces substantial social and political barriers, and is suitable mainly for shorter trips. This chapter explores the role electric two-wheelers (including pedelecs, e-mopeds and e-scooters) can play in urban vehicle ecosystems, using the sustainable mobility paradigm. Compared to traditional transport planning, this paradigm has a stronger focus on aspects such as accessibility, people, streets as a space, city liveability, as well as environmental impacts. The analysis is based on existing literature in the academic and policy realm and a comparison with other transport modes including motorcycles, bicycles, public transport and cars. It includes cases from China, Vietnam and the Netherlands, each of which have distinct mobility system characteristics. Possible policy instruments to facilitate further deployment of electric two-wheelers are discussed as well. This chapter thereby helps filling a gap in transport, sustainable development and climate change mitigation literature, in which electric two-wheelers have not been well covered to date.

Keywords: e-bikes, electric two-wheelers, urban mobility, accessibility, sustainable transport, transport policy

1. Introduction

Substantial changes in urban mobility systems are required across the globe in order to reach sustainable development goals (SDGs) and climate change objectives. Indeed, sustainable urban transport is essential to achieve SDG targets related to urban access, reduced health impacts from air pollution and road traffic crashes, clean energy, inefficiency of fossil fuel subsidies, resilient infrastructure, climate change measures and sustainable cities and communities [1].

Exposure to ambient air pollution results in 4.2 million deaths annually [2], and 9 out of 10 urban dwellers breathe air that does not comply with the WHO standards. The transport sector is the largest contributor to PM2.5 pollution [3], with, for example, the economic cost of air pollution from road transport in OECD countries estimated at close to USD 1 trillion per year, measured in terms of the value of lives lost and ill health [4]. In addition, physical inactivity is responsible for 3.2 million deaths annually [5]. Data on noise impacts are limited; however, the health effects from exposure to noise are substantial. In the European Union, 65% of the urban population is exposed to noise above 55 dB, leading to an estimated 1–2 million DALYs annually [6]. Other concerns related to urban transport include increasing congestion and lack of accessibility, social equity and transport justice [7], urban liveability, habitat fragmentation [8] and energy security or oil price vulnerability [9].

At the global level, urban (passenger) transport contributed about 2 billion tonnes of CO_2 in 2015 or one-quarter of total transport emissions (including international aviation and maritime transport) [4]. In order to meet the Paris Agreement climate change targets of staying well below 2° and aiming for 1.5°, total transport emissions have to be limited to 2–3 billion tonnes in 2050 from approximately 8 billion tonnes today [10]. As transport is where emissions are rising the most rapidly among all sectors, this is a major challenge.

Traditionally, transport planning has been focussing on providing infrastructure to meet a projected future demand for transport. As Banister argued in his seminal article, 'The Sustainable Mobility Paradigm' [11], addressing the above-mentioned challenges requires fundamental shifts in transport planning. This new paradigm involves, among others, focussing on accessibility of opportunities in the urban space rather than mobility and vehicular travel per se, management of travel demand, considering street as a 'space' rather than a road and including environmental and social concerns in transport project evaluation.

Most literature on sustainable mobility considers modal shift away from cars to more environmentally benign modes. Within this, there is a significant and growing amount of research on cycling, with bicycles also enjoying a good reputation with policymakers in most countries. Motorcycles, on the other hand, are, by and large, ignored in sustainable transport research, as well as in policy discussions and climate change literature, although some researches are acknowledging its importance in meeting climate change objectives [12]. Since 2010, more literatures have been published on electric two-wheelers, such as e-bikes and electric scooters, which assess their potential impacts on the mobility systems and the environment, user motivations and aspects such as safety and driver behaviour [13].

This chapter aims to add to existing literature by (1) assessing qualitatively how electric two-wheelers (E2W) can contribute to sustainable urban mobility and (2) reviewing policy options to enhance the role of E2W.

In this chapter, we adopt the following approach. The e-bike is a relatively new vehicle and in fact represents 'the most rapid uptake of alternative fuelled vehicles in the history of motorisation' [13]. E2W thereby change urban mobility ecosystems or regimes, including the vehicle

http://dx.doi.org/10.5772/intechopen.81460

fleet composition, mobility options and urban planning. As a new mobility option for both passenger and freight (e-cargobikes), it has the potential to replace trips by other modes, notably bicycles, motorcycles, public transport and cars but also small trucks and vans. In addition, E2W may support other modes such as public transport by providing convenient first and last mile mobility. We particularly look at shifting from motorcycles to E2W as a way to maintain space-efficient (passenger) mobility on two wheels, yet in a cleaner fashion; however, shifting away from other modes may also be significant.

2. Conventional and electric two-wheelers: current situation

Two-wheelers powered by a combustion engine, including mopeds, scooters and motorcycles, play a key role in transport in many parts of the world. In South and Southeast Asia, the share of two-wheelers in the passenger vehicle population is particularly high, for example, 72% in India [14], 87% in Indonesia and 95% in Vietnam (95%) [15]. The fleet is growing by up to 10% annually in several Asian countries. In Europe, 12% of all registered vehicles are powered two-wheelers, though almost half of these are >250 cc [14] and often used for touring and sports purposes rather than utility mobility. In Brazil, 26% of the vehicle population are motorcycles and in North America, only 3% [14]. In China and Southeast Asia, two-wheelers account for more than a quarter of total passenger transport activity (measured in passenger-km) in 2015, more than any other mode [12].

A motorcycle uses seven times less space compared to a car [16] and is more energy-efficient, even when accounting for a higher average occupancy of cars. Nevertheless, motorcycles are significant sources of air pollution, noise and CO_2 emissions and are involved in a large share of road crashes.

In many cities in Asia, motorcycles are the preferred mode of transport by a large share of the population [17]. In dense cities, for example, Hanoi, accessibility to jobs by motorcycle is higher than by any other modes [18]. It enjoys higher status than a bicycle and is considered more convenient, particularly for hot and humid weather conditions. Motorcycles are relatively affordable and can be acquired by most households, thereby making it an equitable mode of transport.

In European cities, motorcycle mode share is often low, however rising in recent years. Amsterdam is a case in point, with about 2% of trips in 2016 by mopeds and scooters, which is a doubling compared to 2008. Moreover, such two-wheelers are involved in 16% of road crashes [19]. Although there are plans and strong public support to ban mopeds and scooters from using bicycle lanes, as of mid-2018, these are still allowed on all bike lanes and are not required to wear helmets. The users are relatively diverse, that is, they include all age and income groups.

At the same time, electric two-wheelers are gaining importance and already take a significant modal share in some countries in 2015, notably China (7%) and 2–4% in Denmark, the Netherlands and Japan [20]. China dominates E2W sales with about 30 million and a stock of about 250 million [17], followed in sales by Europe with 2.3 million, while the rest of the world accounts for about 1 million in 2015 [13].

These are predominantly e-bikes, which fall roughly in two categories: pedal-assisted bicycles (also called pedelecs) or throttle-controlled electric mopeds with the option of pedal power (often for regulatory purposes). The former are the most common in Europe, while in China and some Southeast Asian countries, e-bikes of the latter types are dominant. The speeds are up to 20–25 km/h for pedelecs and up to 45 km/h for e-mopeds, and the weight ranges from 20 to 45 kg. Electric scooters are capable of higher speeds and often fall in a different vehicle category than e-bikes.

To explore the characteristics of different E2W markets and their role in the mobility system, we examine—briefly and in broad terms—developments in China, Vietnam and the Netherlands.

In China, petrol-fuelled motorcycle bans in many cities in the early 2000s—starting with a sales ban in Shanghai in 1996 [21]—have resulted in a large and fast uptake of especially scooter-style e-bikes, with a 15–25% trip mode share in major cities [22]. They have become popular as well in cities where conventional motorcycles are still allowed (**Figure 1**). E-bikes have pedals and are classified as nonmotorised vehicles, thereby there is no requirement to wear helmet nor for licencing, and they can use bicycle lanes. E-bikes are used by a wide range of user groups, in small, medium and large cities, also as access mode for public transport. However, in general two-wheelers are used more by lower-income groups [23].

Within Southeast Asia, Vietnam has the highest share of motorcycles in the vehicle fleet [15]; however, e-bikes are not yet popular. The main users are those that do not have access to motorcycles. A key user group is students, who find these more convenient than the bicycle while it does not require a licence, registration or helmet, as is the case for motorcycles. E-bikes (**Figure 2**) are often of type that could be considered a hybrid between scooter-style and bicycle-style e-bikes. They have pedals, yet these are rarely used. There are two-wheeler lanes in some cities; however, in general they have to mix with other traffic. There are no specific policies to promote electric two-wheelers. A ban on motorcycles is planned in the city centre of Hanoi in 2030; however, whether this will cover e-bikes is not clear.

In Europe, most e-bikes are of the bicycle style. Weather is less of a barrier to cycling compared to Asian countries; however, e-bike extends the range of trips that can be covered by

Figure 1. Guilin, China: both conventional and electric motorcycles are used.

Figure 2. Hanoi, Vietnam: two-wheelers in mixed traffic, although on some roads space is allocated by mode.

Figure 3. Amsterdam, the Netherlands: bike lanes are used by bicycles, mopeds and electric bikes.

bicycle, in distance and user groups. Looking at the Netherlands, e-bikes sales are increasing rapidly and in 2017 account for about one-third of bicycle sales [24]. Although in the early days of e-bikes it was mostly the elderly buying E2W, in recent years, its popularity is spreading to many other groups, including students, commuters and parents of young children. Approximately half of the km travelled is for recreational purposes [24]. E-bikes, just as mopeds, are allowed on the bicycle lanes (**Figure 3**); however, as of July 2018, speed pedelecs need to use the main road in urban areas.

3. The potential role of electric two-wheelers in sustainable urban mobility

In this section, we review how E2W fits in a sustainable urban mobility system. Based on Banister [11], we focus on aspects of accessibility, as well as social and environmental sustainability.

Most trips in cities across the globe are less than 10 km, and many are shorter than 5 km. In the Netherlands, over 50% of car trips are shorter than 7.5 km [24], and in medium-sized cities such as Rajkot and Visakhapatnam (1–2 million population) in India, nearly 80% of all trips are below 5 km [25]. Even in the United States, known for their relatively low density and sprawl, 35% of all car trips are less than 5 km and 60% are below 8 km [20]. Trips up to 5 km are generally considered to be suitable for cycling. E-bikes can extend the range of trips currently or potentially undertaken by bicycle [24, 26] by reducing barriers such as hilly terrain, weather, low speed, physical strain and bad air quality [13]. At the same time, e-bikes can replace trips by motorised modes such as motorcycle, car and public transport.

Mode shift impacts vary, as shown in emerging literature. In Chinese cities where petrol-fuelled motorcycles are not allowed, e-bikes are displacing bus trips, yet also a significant share of trips by car/taxi and bicycle [13]. Some of the factors that impede use of bicycles are also barriers to using e-bike, for instance, heat, rain and air quality [26]. In Sweden, more than 50% of e-bike users in both urban and rural areas report replacing car trips, across different trip purposes [27]. An analysis of studies in Europe shows that the proportion of e-bike trips that replace car trips ranges from 16 to 76% [28].

Two-wheelers provide accessibility benefits over cars and public transport. Especially in dense cities in Asia and Europe, bicycles and motorcycles are more affordable, flexible, reliable and often faster than cars. In addition to reduced space require for parking, motorcycles use 3.4 times less road space than cars in Hanoi [29]. Two-wheelers are thereby much more space-efficient than other private vehicles, even when considering a slightly higher average occupancy rate for cars. In the longer term, mobility based on two wheels thus has an impact on land use and urban development and enables denser and liveable cities as opposed to 'sprawling car-dominated cities' [20]. Therefore, mobility modes and transport planning are strongly related to more fundamental questions around urban development and the future of cities.

Other positive impacts of e-bikes—compared to car or motorcycle travel—for individuals and society include health due to physical activity [13] and social interaction in the public space [30], although both these effects are less strong than for bicycle travel. In the economic realm, electrification of transport is a key strategy to reduce oil consumption and improve energy supply security.

Environmental benefits are substantial as well, in particular for climate change, air quality and noise. Compared to conventional motorcycles, electric two-wheelers emit substantially less CO_2 emissions per km (on a life cycle basis), even when powered by coal-based electricity [31]. Indeed, over 80% of the 29 million tonnes of CO_2 savings in 2017 by all types of electric vehicles globally are due to e-bikes in China [17]. For Vietnam, e-bikes are identified as the option with the second-largest CO_2 abatement potential in the transport sector [32]. Kerdlap and Gheewala [33] show that in Thailand, deploying a fleet of 13.6 million electric motorcycles to replace an equivalent fleet of conventional petrol-powered motorcycles between 2015 and 2030 could reduce two-wheeler life cycle CO_2-eq emissions by approximately 42–46%. Globally, it is estimated that in 2050, 22% of urban passenger travel can be by (e) bike, compared to 6% in the base case. This results in 300 $MtCO_2$ reductions in 2050 and USD 1 trillion in savings from vehicle purchase and operation and construction and maintenance of infrastructure [20]. Moreover, meeting the Paris Agreement targets requires 70% of global two-wheelers to be electric [17]. It should be noted that in peer-reviewed literature on climate change mitigation, however, there is limited attention for the role of electric two-wheelers.

Urban air pollution impacts are significant as well. The health impact of particulate matter on a person-km basis is lower than other modes that include petrol-powered cars, even for a coal-based electricity grid [34]. Finally, e-bike can reduce noise significantly [35], as conventional motorcycles and cars are key sources of urban traffic noise.

Table 1 presents a qualitative assessment of the sustainability aspects of electric two-wheelers discussed above, in comparison with other modes. It should be noted that this comparison is for illustration purposes only, as the modes cover different trip distances and impacts may differ considerably depending on local circumstances, particularly transport planning and environmental standards. Public transport modes are bus, tram and metro. Paratransit covers a variety of more informally organised transport such as motorcycle taxis, three-wheelers and minibuses, which are sometimes used as feeder mode for high-capacity public transport. Equity assessment is based on typical costs of travel, including ticket prices for public transport and total costs of ownership for private vehicles. In space efficiency, both parking and road space are considered. Lower CO_2 emissions also imply improved energy efficiency (i.e. reduced resource consumption) as well as energy security. Public transport first and last mile trips often involve 'active' modes such as walking or cycling. Other environmental impacts, such as pollution from battery production, are not explicitly considered here.

In the above assessment, road safety has not been taken into account. A meaningful comparison between modes is not possible as often multiple modes are involved in one road crash. In addition, among countries, road crash rates and fatalities for all modes differ by more than an order of magnitude. That said, motorcycles are often associated with safety concerns, and globally 23% of the 1.3 million road deaths were drivers or passengers of motorised two- and three-wheelers [36]. In the Netherlands, an increase in road deaths in 2017 was associated

Mode	Typical trip distance	Equity	Space efficiency	Air pollution	CO_2 emissions/ energy use	Noise	Physical activity
Walking	<1.5	+++	+++	+++	+++	+++	+++
Cycling	1–5	+++	+++	+++	+++	+++	+++
E2W	1–15	++	++	+++	+++	+++	+
Motorcycle	1–15	++	++	+	+	0	0
PT + NMT	1–20+	++	+++	++	++	++	++
PT + paratransit	1–20+	++	+++	+	+	+	+
Paratransit	1–5	++	++	0	+	+	0
Car	1–20+	+	0	+	0	+	0

Notes: PT, public transport; NMT, nonmotorised transport; 0, lowest rating; +, low rating; ++, medium rating; +++, high rating.

Table 1. Environmental and social sustainability impacts of urban transport modes.

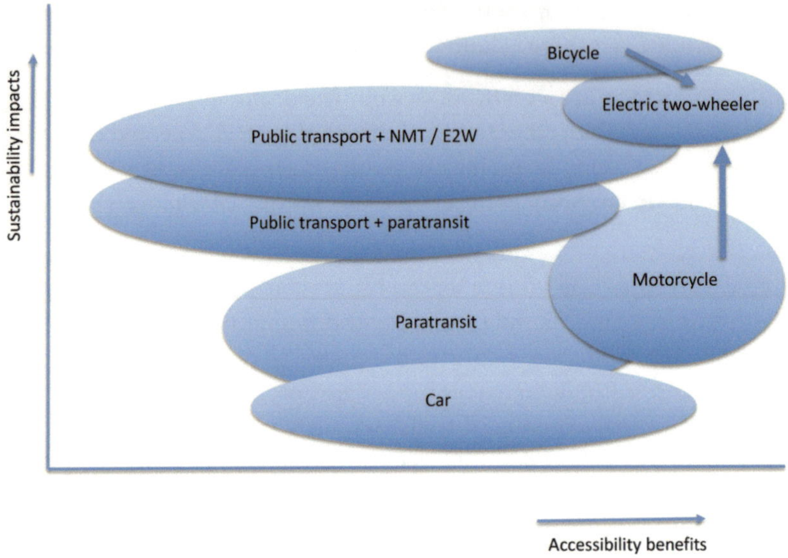

Figure 4. Indicative qualitative assessment of sustainability impacts and accessibility benefits of urban transport modes for trips 2–10 km, on a person-km basis. Accessibility covers travel time including parking and reliability. Sustainability aspects here include equity, road and parking space efficiency, air pollution, CO_2 emissions and fuel consumption, noise and physical activity (see **Table 1**). Larger ovals indicate larger spread in accessibility/sustainability benefits depending on local conditions. All vehicles except E2W are powered by internal combustion engines.

with the rising use of e-bikes among the elderly, as these are more difficult to handle than bicycles [37], with e-bikes taking 30% of total bike fatalities [24]. Unruly driving behaviour, for example, jumping red lights at intersections, is a key concern with e-bike riders as well as bicyclists and motorcycles in various countries [13].

In **Figure 4**, the sustainability impacts as included in **Table 1** are added and compared with accessibility benefits for urban trips of 2–10 km length in a dense city that has a balanced approach to transport planning for the different modes. Accessibility, or the ease of reaching opportunities, covers travel time, flexibility, reliability and ease of parking. The width of the ovals indicates how strongly accessibility depends on local conditions such as congestion, parking options, public transport service quality and urban planning. For example, in some cases a car may be as fast as a motorcycle, whereas in heavy congestion and limited parking availability, a car provides low accessibility.

A key observation from this figure is that E2Ws increase range and comfort of bicycles and improve sustainability performance of motorcycles while preserving accessibility benefits.

4. Policy options

In various countries, a limited number of policy initiatives are taken to promote electric mobility on two wheels. Yet in general, it can be stated there is a lack of policy attention, particularly in comparison to policy and research on other electric vehicles such as cars and buses.

There is a large potential in different regions to expand the use of electric two-wheelers, so what can be done do to harness this potential? Policymakers have a range of instruments that can be deployed [17]. In general, these can be organised by regulatory, economic and informative instruments [38], while for transport, often planning instruments are considered as well [39]. **Table 2** presents a brief overview, after which these options are discussed.

In the realm of regulatory instruments, the strongest policy measure is to ban motorcycles powered by fossil fuels, as implemented in Chinese cities. A phase out of conventional motorcycles and moped sales is considered in the Netherlands [40]. Similarly, a low-emission zone in a city can be designed such that conventional motorcycles will not be able to comply with the required emission standard to be allowed to circulate in the zone. Further, to improve safety for two-wheelers, speed limits for shared roads can be an effective tool, e.g. 30 km/h in urban areas where no dedicated lanes exist. At the same time, helmet use can be made compulsory for vehicles capable of travelling faster than a certain speed, for example, 25 km/h. Finally, electric two-wheelers need an appropriate legal framework in national vehicle legislation. Malaysia, for example, has adopted standards for electric mopeds with speeds in the range of 25–50 km/h, covering safety, performance and national compliance issues [41].

Planning instruments are key as well, for example, allocating dedicated road space for two-wheelers, together with quality standards for existing and new road surface that improves safety [14]. E2W then co-exist with either conventional motorcycles or bicycles, depending on the desired speed range and design of the two-wheeler lanes. For example, in China, e-bikes are often allowed on bike lanes. In addition, two-wheeler mobility can be made

Instrument type	Policy measure
Regulatory instruments	Phase-out conventional motorcycles
	Low-emission zones in cities
	Speed limit 30 km/h on shared roads
	Vehicle standards and registration requirement
Planning instruments	Dedicated lanes for E2W (and bicycles)
	Travel demand management, including traffic calming and parking management
	Dedicated waiting boxes at intersections, optionally with shading
	Electric bike-sharing facilities
Economic instruments	Incentives such as subsidies, tax breaks for purchase or registration of E2W
	Taxation of fuels (petrol and diesel)
Information/communication instruments	Campaigns
	Behaviour change programmes

Table 2. Overview of policy options to promote E2W.

more attractive, for example, by advanced stop boxes for two-wheelers at intersections (see **Figure 1**, right-hand side), which can be shaded in tropical regions to protect drivers from heat and rain. Travel demand management, to make car travel less attractive, is required as well. Measures include traffic calming (e.g. speed bumps), reducing parking supply, fuel taxation and restricting access to roads while allowing two-wheelers, ensuring a lower 'detour' factor for the latter. Finally, electric bikes can be included in bike-sharing schemes.

Financial incentives are another important instrument to promote E2W purchase and use. These can be designed, for example, as purchase subsidies, as done in some provinces in China [21], sales tax breaks [42] or an increase in petrol tax. In countries where motorcycle taxis are common, specific programmes to convert motorcycle fleets in a city to electric two-wheelers can be designed. In the Netherlands, a tax-deduction scheme for employees to buy a bicycle or e-bike exists. For behaviour change policies, incentives are often used in tandem with information instruments. It is argued that key life-changing events, such as moving to a new city or the birth of a child, are often powerful catalysts for behaviour change. Policy instruments can mimic such disruptive change [43], e.g. by pilot programmes with free e-bikes in exchange for car keys, as done in a promotion programme in Switzerland [44] or other incentives and campaigns. At the same time, information and training on road safety are needed for drivers of two-wheelers and other road users, e.g. as part of driver training curricula [14].

5. Conclusions

Electric two-wheelers can play an important role in sustainable urban mobility systems and addressing climate change. However, two-wheelers, including electric, suffer from a bad reputation and a lack of attention from policymakers in many countries. This chapter gives an overview of developments in China, Vietnam and the Netherlands, each with their own mobility system characteristics. Other than China, e-bikes and electric scooters are still in an early stage of development.

Yet there is a large potential: electric two-wheelers can on the one hand address negative impacts of fossil-fuelled motorcycles and cars on air quality, climate and noise while on the other hand extend the distance range of bicycles, by reducing the physical effort needed, which is especially attractive in hilly and tropical environments. In general, compared to other modes, electric two-wheelers score high on key criteria for sustainable mobility in terms of accessibility (flexibility, reliability, and speed), road space use, equity and environmental externalities, although road safety remains a concern.

Finally, this chapter proposes a range of policy measures—regulatory, planning, economic and communicative instruments—that can be used to promote purchase and use of electric two-wheelers. These include for example implementing low-emission zones, phasing out of conventional motorcycles, improving the legal framework, urban planning to increase attractive and safety of two-wheelers and conducting behaviour change programmes that cover both incentives and information for individual users or motorcycle taxi fleets.

Acknowledgements

Comments and feedback from the book editor, Kathleen Dematera and Todd Litman, on a draft version of this chapter are greatly appreciated.

Author details

Stefan Bakker

Address all correspondence to: sjabakker@gmail.com

Independent Consultant, Amsterdam, The Netherlands

References

[1] Sustainable Mobility for All initiative. Global Mobility Report 2017. Tracking Sector Performance. Available from: http://sum4all.org/publications/global-mobility-report-2017 [Accessed: 9 September, 2018]

[2] World Health Organisation. Ambient (Outdoor) Air Quality and Health. Key Facts. Available from: http://www.who.int/en/news-room/fact-sheets/detail/ambient-(outdoor)-air-quality-and-health [Accessed: 9 September 2018]

[3] Karagulian F, Belis CA, Dora CF. Contributions to cities' ambient particulate matter (PM): A systematic review of local source contributions at global level. Atmospheric Environment. 2015;**120**:475-483. DOI: 10.1016/j.atmosenv.2015.08.087

[4] International Transport Forum: Transport Outlook. Paris: OECD Publishing; 2017

[5] World Health Organization. Physical Activity for Health. More Active People for a Healthier World: Draft Global Action Plan on Physical Activity 2018-2030. WHO Discussion Paper. 9 April 2018. Available from: http://www.who.int/ncds/governance/who-discussion-paper-gappa-9april2018.pdf?ua=1 [Accessed: 9 September, 2018]

[6] European Environment Agency. Environmental Noise. Available from: https://www.eea.europa.eu/airs/2017/environment-and-health/environmental-noise [Accessed: 9 September, 2018]

[7] Martens K. Transport Justice. Designing fair transportation systems. New York and London: Routledge; 2017

[8] European Environment Agency: Landscape Fragmentation Pressure From Urban and Transport Infrastructure Expansion. Available from: https://www.eea.europa.eu/data-and-maps/indicators/mobility-and-urbanisation-pressure-on-ecosystems/assessment [Accessed: 9 September, 2018]

[9] Leung A, Burke M, Yen B, Chiou Y-C. Benchmarking urban transport oil vulnerability in 11 Asia-Pacific cities. Journal of the Eastern Asia Society for Transportation Studies. 2017;**12**:1005-1022

[10] Gota S, Huizenga C, Peet K, Medimorec N, Bakker S. Decarbonising transport to achieve Paris Agreement targets. Energy Efficiency. 2018:1-24. DOI: 10.1007/s12053-018-9671-3

[11] Banister D. The sustainable mobility paradigm. Transport Policy. 2008;**15**:73-80. DOI: 10.1016/j.tranpol.2007.10.005. https://link.springer.com/article/10.1007/s12053-018-9671-3

[12] International Energy Agency. Energy Technology Perspectives. Paris: OECD Publishing; 2016

[13] Fishmann E, Cherry C. E-bikes in the mainstream: Reviewing a decade of research. Transport Reviews. 2016;**36**:72-91. DOI: 10.1080/01441647.2015.1069907

[14] International Motorcycle Manufacturers Association: The Shared Road to Safety. A Global Approach for Safer Motorcycling. 2016. Available from: http://immamotorcycles.org/sites/default/files/_ftp-pdfs/2016-IMMA_Webversion.pdf [Accessed: 9 September 2018]

[15] Bakker S, Dematera K, Kappiantari M, Nguyen AT, Guillen MC, Gunthawong G, et al. Low-carbon transport policy in four ASEAN countries: Developments in Indonesia, the Philippines, Thailand and Vietnam. Sustainability. 2017;**9**:1217-1233. DOI: 10.3390/su9071217

[16] The World Bank: Motorization and Urban Transport in East Asia. Motor Scooter & Motorbike Ownership & Use in Hanoi. Final report. 2015

[17] International Energy Agency. Global EV Outlook 2018. Towards Cross-modal Electrification. Paris: OECD Publishing; 2018

[18] Nguyen NQ, Zuidgeest M, Van den Bosch F, Sliuzas RV, Van Maarseveen F. Using accessibility indicators to investigate urban growth and motorcycle use in Ha Noi city, Vietnam. Proceedings of the Eastern Asia Society for Transportation Studies. 2013;**9**. http://easts.info/on-line/proceedings/vol9/PDF/P126.pdf

[19] Gemeente Amsterdam. Meerjaren Plan Verkeersveiligheid 2016-2021 (in Dutch). Available from: https://www.amsterdam.nl/parkeren-verkeer/verkeersveiligheid/mjp-verkeersveilig/ [Accessed: 9 September, 2018]

[20] Mason J, Fulton L, McDonald Z. A Global High Shift Cycling Scenario: The Potential for Dramatically Increasing Bicycle and E-Bike Use in Cities Around the World. California: ITDP/UC Davis; 2015

[21] Ruan Y, Hang CC, Wang YM. Government's role in disruptive innovation and industry emergence: The case of the electric bike in China. Technovation. 2014;**34**:785-796. DOI: 10.1016/j.technovation.2014.09.003

[22] Wang J. Electric Two-wheelers in China. In: Presentation at ASEAN workshop on two-wheelers, Manila, November 2017. Available from: https://www.transportandclimatechange.org/download/tuewas-unep-workshop-on-two-wheelers-country-input-china-by-wang/ [Accessed: 9 September, 2008]

[23] Pan H, He X, Wang L. Mobility Improvements by electric two-wheels in public transit under-developed area. In: World Conference on Transport Research—WCTR 2016 Shanghai; 10-15 July 2016; Transport Research Procedia. 2017

[24] Harmsen L, Kansen M. Fietsfeiten. Den Haag, The Netherlands: Kennisinstituut voor Mobiliteit; 2016. Available from: https://www.kimnet.nl/mobiliteitsbeeld [Accessed: 9 September, 2018]

[25] Tiwari G, Jain D, Rao KR. Impact of public transport and non-motorized transport infrastructure on travel mode shares, energy, emissions and safety: Case of Indian cities. Transportation Research Part D. 2016;**44**:277-291. DOI: 10.1016/j.trd.2015.11.004

[26] Campbell A, Cherry C, Syerson M, Yang X. Factors influencing the choice of shared bicycles and shared electric bikes in Beijing. Transportation Research Part C. 2016;**67**:399-414. DOI: 10.1016/j.trc.2016.03.004

[27] Winslot Hiselius L, Svennson Å. E-bike use in Sweden—CO_2 effects due to modal change and municipal promotion strategies. Journal of Cleaner Production. 2017;**141**:818-824. DOI: 10.1016/j.jclepro.2016.09.141

[28] Cairns S, Behrendt F, Raffo D, Beaumont C. Electrically-assisted bikes: Potential impacts on travel behaviour. Transportation Research Part A. 2017;**103**:327-342. DOI: 10.1016/j.tra.2017.03.007

[29] Nguyen CY, Sano K, Tran VT, Nguyen VA. Estimating capacity and vehicle equivalent unit by motorcycles at road segment in urban road. Journal of Transportation Engineering. 2012;**138**:776-785. DOI: 10.1061/(ASCE)TE.1943-5436.0000382

[30] Brommelstroët M, Nikolaeva A, Glaser M, Skou Nicolaisen M, Chan C. Travelling together alone and alone together: Mobility and potential exposure to diversity social interaction. Applied Mobilities. 2017;**2**:1-15. DOI: 10.1080/23800127.2017.1283122

[31] International Energy Agency. Global EV Outlook. Paris: OECD Publishing; 2016

[32] Asian Development Bank. Pathways to low-carbon development for Viet Nam. Asian Development Bank. 2017. Mandaluong City, the Philippines. Available from: https://www.adb.org/sites/default/files/publication/389826/pathways-low-carbon-devt-viet-nam.pdf [Accessed: 9 September, 2018]

[33] Kerdlap P, Gheewala S. Electric motorcycles in Thailand. A lifecycle perspective. Journal of Industrial Ecology. 2016;**20**:1399-1411. DOI: 10.1111/jiec.12406

[34] Ji S, Cherry C, Bechle M, Wu Y, Marshall J. Electric vehicles in China: Emissions and health impacts. Environmental Science & Technology. 2011;**46**:2018-2024

[35] Sheng N, Zhou X, Zhou Y. Environmental impact of electric motorcycles: Evidence from traffic noise assessment by a building-based data mining technique. Science of the Total Environment. 2016;**554-555**:73-82. DOI: 10.1016/j.scitotenv.2016.02.148

[36] World Health Organization. Global Status Report on Road Safety. Geneva: WHO Press; 2015

[37] Stichting Wetenschappelijk Onderzoek Verkeersveiligheid. Verkeersdoden Nederland (Traffic fatalities Netherlands, in Dutch). Factsheet. Available from: https://www.swov.nl/feiten-cijfers/factsheet/verkeersdoden-nederland [Accessed: 9 September, 2018]

[38] Givoni M, Macmillen J, Banister D, Feitelson E. From policy measures to policy packages. Transport Reviews. 2013;**33**:1-20. DOI: 10.1080/01441647.2012.744779

[39] Wittneben B, Bongardt D, Dalkmann H, Sterk W, Baatz C. Integrating sustainable transport measures into the clean development mechanism. Transport Reviews. 2009;**29**: 91-113. DOI: 10.1080/01441640802133494

[40] Tweede Kamer der Staten-Generaal (Dutch Parliament). Motie van de leden Dik-Faber en Van Tongeren. Tweede Kamer, vergaderjaar 2016-2017, 34 550 XII, nr. 49

[41] Malaysian Standards. Electric mopeds—Specification. MS Standard 2688. 2018. Available from: https://www.transportandclimatechange.org/wp-content/uploads/2018/07/MS-2688-Preview-1.pdf [Accessed: 9 September, 2018]

[42] Jones L, Cherry C, Vu TA, Nguyen QN. The effect of incentives and technology on the adoption of electric motorcycles: A stated choice experiment in Vietnam. Transportation Research Part A. 2013;**57**:1-11. DOI: 10.1016/j.tra.2013.09.003

[43] Plazier P. Power to the pedals. Perspectives on the potential of e-bike mobility for sustainable and active transport systems [thesis]. Groningen, University of Groningen; 2018

[44] Moser C, Blumer Y, Hille SL. E-bike trials' potential to promote sustained changes in car owners mobility habits. Environmental Research Letters. 2018;**13**:044025. DOI: 10.1088/1748-9326/aaad73

Chapter 7

Influence of the Motor Transport on Sustainable Development of Smart Cities

Irina Makarova, Ksenia Shubenkova,
Vadim Mavrin and Eduard Mukhametdinov

Additional information is available at the end of the chapter

http://dx.doi.org/10.5772/intechopen.71045

Abstract

The transport system is one of the fundamental intelligent systems in the Smart City, and one of the main directions to ensure sustainability and safety of the city transport system is the concept of smart vehicles. Herewith, all processes at all stages of the life cycle should be intellectualized. Since the production stage of the life cycle is one of the most important, the introduction of smart technologies (Industry 4.0) in automotive industry will allow not only to optimize the processes and improve product quality but also to establish favorable conditions for the subsequent intellectualization of the automotive service. The benefits of using smart transport in all fields of activities as well as intellectualization of the decision-making process by the example of the automotive industry enterprises are presented in this chapter.

Keywords: intellectualization, Smart City, industry 4.0, life cycle, automotive industry

1. Introduction

Automotive industry belongs to those sectors of economy, which largely determine the development of other industries where automotive equipment is used, because automotive vehicles can help to solve the problem of population mobility and carry out door-to-door cargo deliveries. High level of motorization and market globalization makes manufacturers to look for new solutions and constantly to improve both the car's design and production technology. Because it is possible to sustain significant competition in the markets only by the continuous development and application of innovative solutions.

The large companies should pay close attention to realization of two main tendencies. The first is a global tendency of a sustainable development, which includes stability of economy, environment and the social sphere. The second is transition to "green economy," which is defined as low carbon, resource efficient and socially inclusive. These two tendencies can be realized by means of rational regulation of the physical, natural and human capital. Therefore, when developing new projects and technologies, it is necessary to consider social consequences of their realization. Potential economic effects of new machinery, and even entire production lines, can only materialize in the case of social efficiency and optimal interaction between man and technology. Moreover, with increasing complication of technical systems, which are becoming more intellectual, the probabilities of failure in such systems are increasingly dependent on erroneous human action. Therefore, the social responsibility of people, who design, create, operate and maintain complex intelligent engineering systems, increases. Thus, it is important to understand the interconnection between all stages of the life cycle of a complex engineering system and to develop management considering this.

2. Smart city as the main direction of urban lands development

2.1. Intellectualization of the complex organizational and technical systems' management

Intellectualization is currently the main trend of the economic and social development. This concept involves a reasonable and rational management and development of all fields of activities. Modern human civilization entered the third millennium and faced with global challenges. The need to solve these problems is formulated in "Millennium Development Goals."

Urbanization is one of the causes of most problems of our millennium. Today, there are 7.3 billion people all over the world, 54% of them live in urban areas. The world has experienced unprecedented urban growth in recent decades. As the population increases, more people will live in large cities. Many people will live in the growing number of cities with over 10 million inhabitants, known as megacities. Different organizations predict [1, 2] that the world population will reach 8.5 billion by 2030 and 27 megacities will exist that time. Analysts also say that there will be 9.7 billion people, and 66% of them will live in urban areas by 2050, with rapid urbanization of the less developed countries.

In information note, Achim Steiner (Executive Director, United Nations Environment Programme) [3] summarized and presented the key findings and policy messages stemming from the Global Environment Outlook (GEO-6) assessments conducted for the six United Nations Environment Programme regions. Each of these regional assessments includes: (1) a review of regional priorities, (2) the state of the environment in the region and the main trends that can affect it in the future and (3) an analysis of the actions so region could become more sustainable. Poor air quality, climate change, unhealthy lifestyles and the disconnection between society and natural environments increasingly affect human health and give rise to new risks. Living within planetary boundaries will require fundamental transitions in energy, food, mobility and urban systems. Transition to an inclusive green economy should be based on viable ecosystems, cleaner production and healthy consumer preferences. There is no doubt that achieving a healthy planet and healthy people requires urgent transformation of the current systems of production and consumption that most contribute to environmental degradation and inequalities in human health and well-being.

At the same time, the development of technique and technologies provided the opportunity of quality transition not only in industry and economy but also in other spheres, including social. Innovation is not only about technology and new ways to deliver services but also about new ways of thinking and finding new opportunities for development. In this case, an important thing is the transition of the control to a qualitatively new level, which enables harmonization of all activity areas within the city and getting the synergies of such interaction.

According to John Wilmoth Director of UN DESA's Population Division [4], "Managing urban areas has become one of the most important development challenges of the 21st century. Our success or failure in building sustainable cities will be a major factor in the success of the post-2015 UN development agenda."

Since ancient times, cities are centers of ideas, culture and science. However, there are many urban problems (including congestion, pollution, noise, diseases, straining land and resources), which should be solved in such a way that allow people to develop socially and economically. This can be achieved in the case of rational urban planning because the high density of cities can increase efficiency and bring technological innovation while reducing resource and energy consumption.

Mobility is a key dynamic of urbanization, and the associated infrastructure invariably shapes the urban environment: the roads, transport systems, spaces and architectural solutions. By 2005, approximately 7.5 billion trips were made in cities worldwide each day. In 2050, there may be three to four times as many passenger-kilometers traveled as in the year 2000. At the same time, the freight turnover may also increase by more than three times.

Nowadays, due to urban sprawl, the distances between points, which generate and attract the passenger flows, have become longer, which leads to the greater dependence of citizens on individual motorized transport. Thus, traffic jams, congestion, pollution, noise stress and traffic accidents are reality of today's megalopolises.

2.2. Smart City and smart mobility

Innovation in transportation today is a very relevant topic [5–10]. More than ever before, we understand that transportation has a key influence on how societies form and develop over time. This is reflected in the concept of Smart City. Herewith, smart mobility is one of the major issues, because it ensures accessibility of workplaces and recreations. Moreover, smart mobility is also a part of production and other subsystems of the city economy.

If we consider Smart Cities from the point of view of economic branches, then it is possible to allocate smart energy, smart transport, smart construction, smart industry and so on. On the other hand, Smart Cities can be considered from the point of view of different city subsystems' objects: smart infrastructure, smart buildings, smart vehicles, etc. And, finally, technologies that improve people's lives (education, medicine, service) can also be "smart"

(**Figure 1**). Whatever way of smart technologies' classification is used, it is impossible to organize processes of any area of economics without transport. It largely relates to the motor transport, because only it can organize door-to-door transportation of cargos and passengers. Therefore, smart technologies that provide design, creation and operation of smart transport in an optimal way and with minimal negative impact on environment and human are very important in transport sphere.

2.3. Smart transport as one of Smart City's development drivers

There is a need for substantial changes in Europe's transport systems, as well as in the mobility behavior of people and businesses in urban areas. Addressing the mobility challenge calls for a paradigm shift in urban planning, encouraging compact cities as a way to increase accessibility and to reduce the need for transportation altogether.

To ensure population mobility means to provide access to all functional destinations, services, places of work, etc. At the same time, city residents should be able to address their needs using as little travel as possible. It can be completed in two ways: (1) reducing the needs to travel by implementing modern information and communications technologies (Internet of Things, Industry 4.0 and other concepts) and (2) reducing distances between places of residence and functional endpoints (the reasons for travel), so that the population could use more sustainable modes of transport, such as walking, cycling, etc.

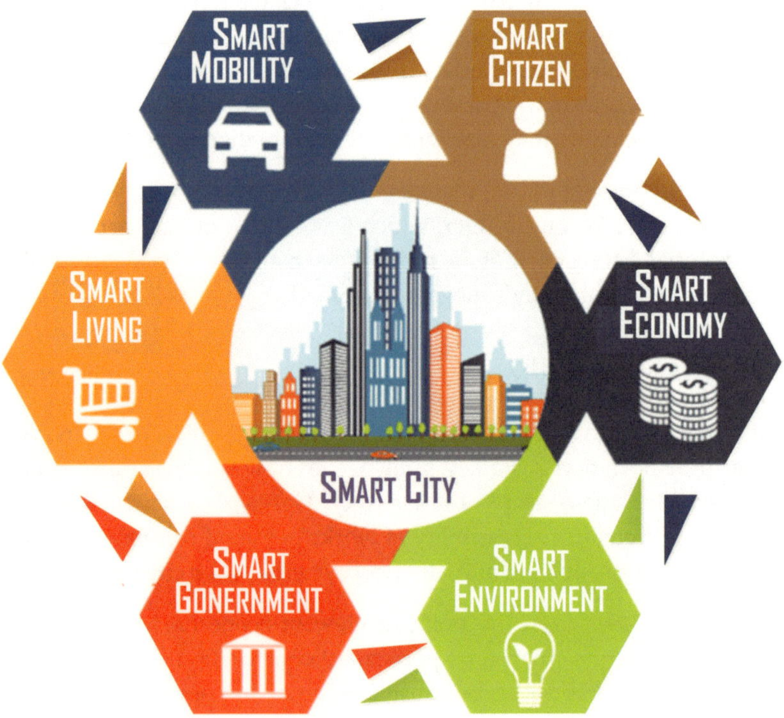

Figure 1. Directions of intellectualization in Smart City.

To make people use more sustainable mobility concepts, it is necessary to ensure possibility to reach any point of passengers' attraction by public transport. However, sometimes it is rather difficult to allow residents easy access to the public transport system; it is the so-called Last Mile Problem. To solve this problem, cities need to provide multi-modal transport systems. For example, bicycle sharing systems can serve as a good way to connect users to public transit networks.

Transport system is one of the major intellectual systems in the Smart City. To ensure its sustainability and safety, the work is being done in three ways: smart infrastructure, smart vehicles and smart users (**Figure 2**). Solutions concern the creation of an efficient and integrated mobility system that allows for organizing and monitoring seamless transport across different modes, increasing the use of environmentally friendly alternative fuels and creating new opportunities for collective mobility.

One of the main areas of ITS, which is actively promoted over the past 15 years, is the implementation of intellectual vehicle. International program "Increased safety vehicle" is implemented. The first experiments of usage of onboard intelligent systems have shown that they are able to reduce the number of traffic accidents by 40% and to reduce

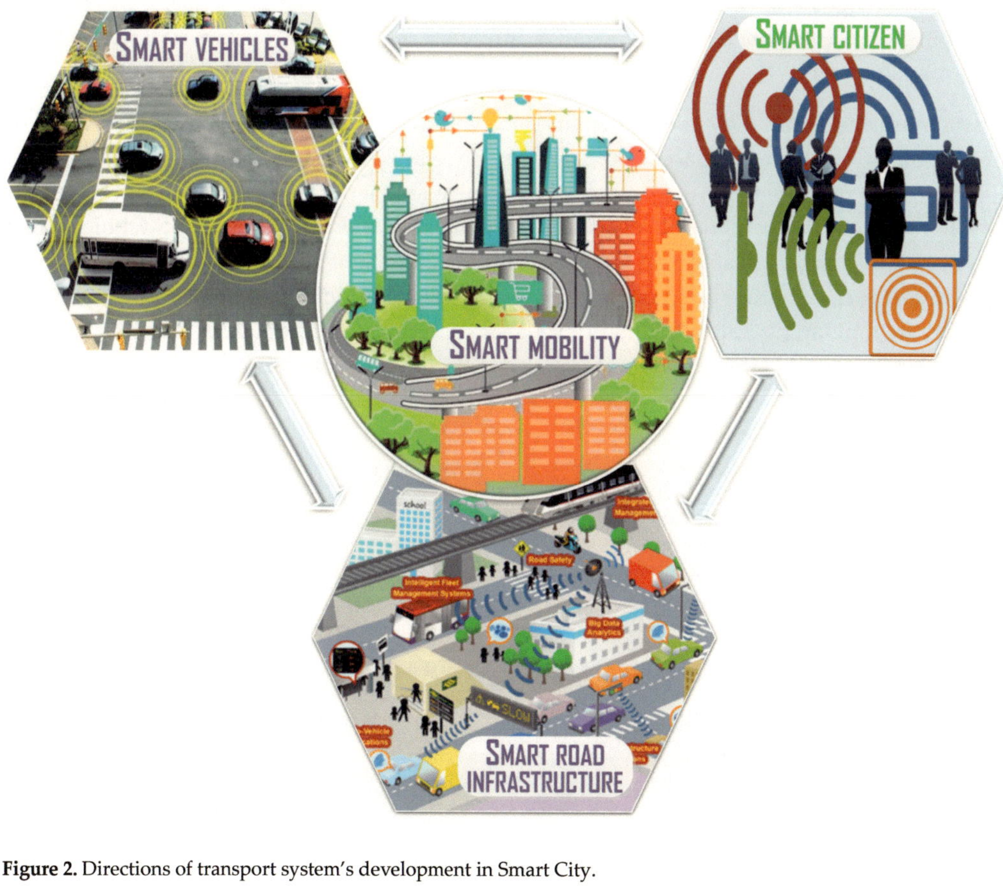

Figure 2. Directions of transport system's development in Smart City.

the number of fatal accidents by 50%. The transition from the creation of driver assistance systems to the development of semi-autonomous unmanned vehicles is a global trend, and it is explained by the desire of developers to ensure the sustainability and the safety of the transport system [11].

However, it should be understood that the emergence of new types of vehicles with fundamentally new control systems could cause problems of security and interaction with other road users. It is especially true in connection with the development of the "livable cities" concept that is aimed at encouraging the use of non-motorized transport, such as walking or cycling. On the one hand, streets need to be adapted, with safe walkways, crossings and cycling lanes, as well as transport junctions need to be established to create safe connection points between different transport modes. On the other hand, it is necessary to identify potential risks of the use of autonomous vehicles, to predict the likelihood of the traffic conflicts (between autonomous vehicles and pedestrians and cyclists, first of all) and to determine the possible consequences. In addition, the ways to prevent risk situations and to reduce the severity of the consequences in case of risk situations should be developed.

Automobile mode of transport is the main one in urban lands, and in the case of unreasonable transport management, it can cause significant problems for other road users. In addition, road transport is the main source of negative influence on the environment, so it needs qualitative management.

The main idea of Smart City is that the city can be "smart" only if the management of all its subsystems is built according to the same rules. If we talk about road transport, then it actually means the management of the vehicle's life cycle as a separate component of the vehicle fleet (**Figure 3**), and at a higher level—the management of the vehicle fleet as a whole. Along with it, all processes at all stages of the life cycle should be intellectualized. At the same time, the orientation to customer needs should be one of the main factors that should be taken into account when planning and implementing these processes. The main directions are creation of the elemental base of intelligent systems and software development.

Figure 3. Negative impact on the environment throughout the life cycle of the vehicle.

3. Industry 4.0 and its role in implementation of the Smart City concept

3.1. Industry 4.0 as the fourth industrial revolution and the prospect for sustainable development of automobile industry

The current state of technics and technologies allows us to create tools and methods not only for managing technical and organizational and technical systems but also devices for analyzing the state of human functional systems and affecting them. This makes it possible to correct and optimize human activity both indirectly, using the recommended loads and parameters, and in real time, which allows creating a comfortable working environment, as well as increasing the safety and efficiency of labor, increasing the efficiency of production systems and product quality.

Real and virtual worlds are now beginning to merge in production that is why we are talking about "Industry 4.0"—the Siemens term for fourth Industrial Revolution. Increasing digitalization and networking is changing the entire industrial production chain, and the volume of data worldwide is exploding. Before analyzing and using the huge amount of data, systems that enable us to understand their content have to be developed. The first step is to get knowledge on what kinds of sensor and measurement technology can be used to collect necessary data and to understand operational principles of systems and devices.

The implementation of the concept Industry 4.0 (**Figure 4**) provides for the formation of cyber-physical systems (CPS), where all elements of the system are active objects that are involved in the exchange of information and make appropriate decisions. Continuous interchange of information in such cyber-physical systems is realized between its elements through the Internet of Things.

The Road Map developed by the group "TechNet" [12] provides creation of new generation of the modern digital productions—"Factory of the future" (**Figure 5**) that is a completely new production environment that is formed by the network of people, things and machines connected to each other. The proposed strategy is based on assumption that replication and scaling of advanced production technologies will determine further development.

Implementation of the "Factory of the future" concept will provide a significant reduction of the time placing on the market of the highly intelligent products by using digital design technologies throughout their life cycle.

Industry 4.0 is aimed at the process optimization, because it covers the entire life cycle, that is, each manufacturer is responsible for his product from the beginning of design and development to disposal.

Classical methods of production organization mean that the flow method can be used only for large quantities of goods. Thanks to the new principles of production processes organization, it becomes possible to manufacture also single products in an industrial way. Industry 4.0, thanks to its flexibility and adaptability provided by cyber-physical systems, can help to realize the mass production of individual orders, which will reduce the price of the product.

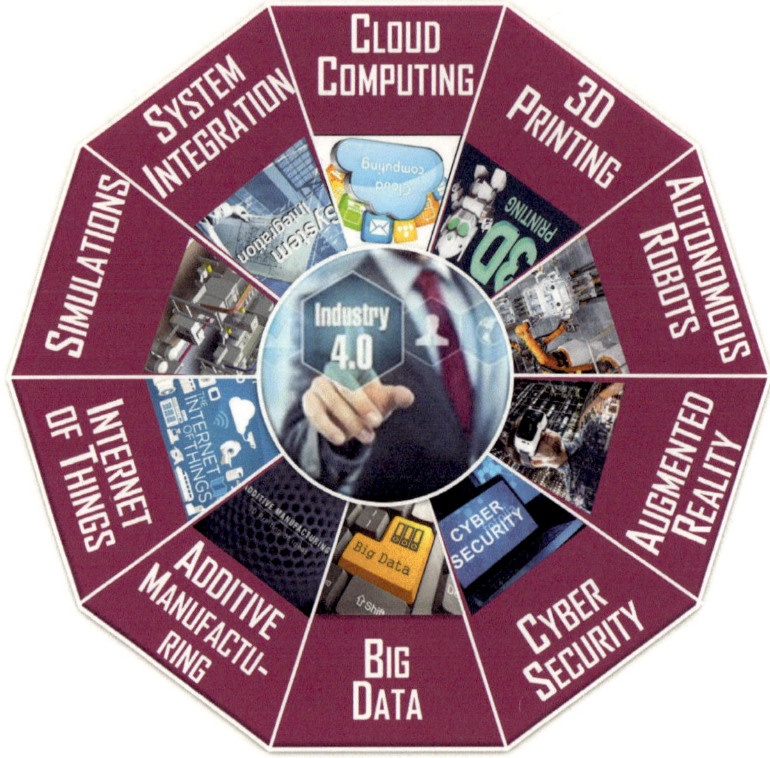

Figure 4. The technologies to implement the concept Industry 4.0.

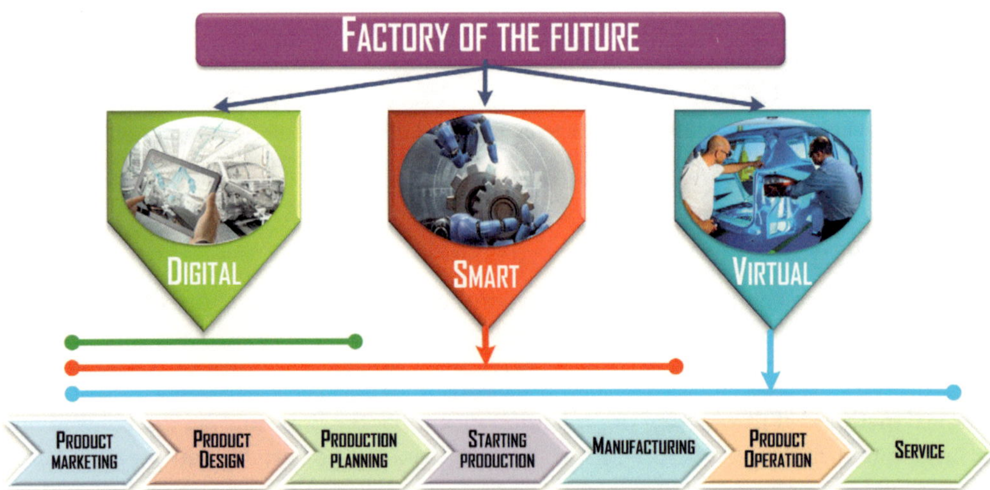

Figure 5. Factory of the future and the product's life cycle.

For production, the ability of various components to communicate through the network opens incredible prospects. In "smart factories," machines will understand their environment and will be able to communicate on a single network protocol among themselves, as well as with the logistics and business systems of suppliers and consumers. The production equipment, receiving information about the changed requirements, will be able to make adjustments to the technological process. As a result, production systems will become capable of self-optimization and self-configuration, the equipment will perform self-diagnostics and further flexibility and individualization of products will occur.

Chrysler's plant in Toledo is an example of the application of cyber-physical systems in manufacturing. Every day it produces more than 700 Jeep Wrangler's bodies. This involves 259 German robots KUKA, which "communicate" with 60,000 other devices and machines. Data interchange and its storage are organized with the use of cloud computing. Modern solutions have significantly increased productivity and flexibility of the factory.

This will also cause the change of the service concept, because the manufacturer will be interested in creating a branded service network that will provide him with implementation of the principle of responsibility for his product throughout the life cycle—from design to disposal (**Figure 6**). This is especially true for modern trucks, which, in contrast to cars, are almost impossible to service in small auto repair workshops. In addition, thanks to the availability of its own service system, the manufacturer will have all information on the features of operation, maintenance and repair of a particular car and the whole park.

3.2. The scope and means of implementing the smart industry concept

The production stage of the life cycle is one of the most important, because exactly at this stage ideas and projects turn into finished products. Besides, the quality of the product depends on the quality of manufacturing. It means that at this stage, it is determined if the targeted audience is large enough, if the product is competitive in the market and how effective and safe are the stages of operation and service.

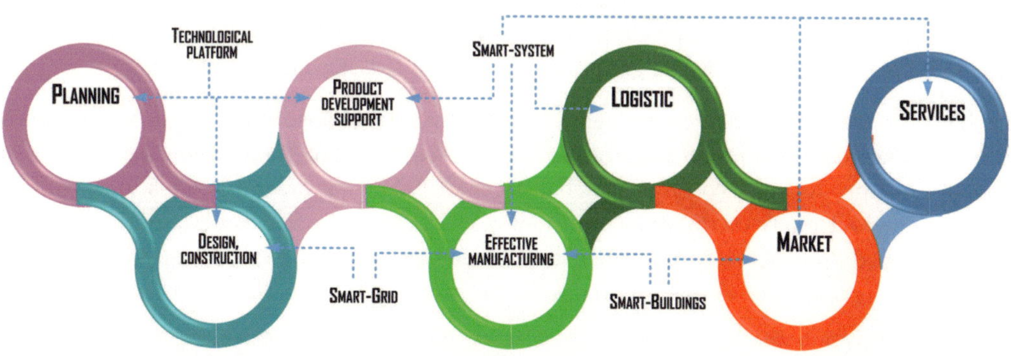

Figure 6. The use of smart technologies at the life-cycle stages.

The terms "Smart Factory," "Smart Manufacturing," "Intelligent Factory" and "Factory of the Future" all describe a vision of what industrial production will look like in the future. Digital technologies will make factories more efficient, intelligent, flexible and dynamic. In connected Industry, everything from design to manufacturing is done through interaction between products and machines and collaborative effort between machines themselves.

Manufacturing in a Smart Factory will be more intelligent, flexible and dynamic in comparison with today's industries. It is so because all production processes and functions (product development; resource planning; logistics; factory and production planning and executing; monitoring, control and management functions, etc.) will be closely interconnected. At the same time, machinery and equipment will have the ability to improve processes through self-optimization and autonomous decision-making.

3.3. Examples of the processes control intellectualization in production systems

3.3.1. Methods and models to improve the assembly line production

In the production process of products with the great amount of components, several problems may occur: stock storage limit, limits of functional zoning, creation of a balanced flow, minimizing the component delivery time to the assembly line positions, etc. Nowadays, in an open and competitive market, companies cannot afford to waste time and resources for work that can be done in a better and faster way with advanced solutions. In Ref. [13], implementing production monitoring systems (PMS) in order to support product lifecycle management (PLM) system with historic knowledge regarding the state of machinery, correctness of assembly operations, etc. is suggested. A module to analyze collected information and predict the future performance of the monitored component is thus needed. In numerous studies, the use of models for this purpose and simultaneous reduction of costs that arise at different stages of production and technological process are suggested. Hence, one of the most important tasks of the organization of the assembly line production is line balancing problem. To balance unilateral and bilateral flow lines, in Ref. [14], a model that takes into account zoning and priority limits, synchronous and positional constraints, buffer time have been developed. The objective function of the model maximizes line efficiency and minimizes index of smoothness and total cost per unit of product. As an example, bilateral assembly line of chassis production of motor vehicles lowering the cost per unit by 42% has been provided.

In the market of automotive components, competitiveness is ensured by high-quality products and low cost. This requires manufacturers to search for methods to minimize costs at all stages of production. To meet the challenges of balancing assembly lines with considering costs exact methods, heuristic and metaheuristic approaches can be applied [15–17]. The developed approaches combine heuristic models and exact algorithms based on "taboo" search in order to minimize short-term operating costs, capital investments, costs of labor and work in progress.

Inefficient production and suboptimal in-plant logistics contribute significantly to environmental degradation. Hence, large number of studies aiming at optimizing the planning of technological transport in view of its negative impact on the environment. Thus, a concept to optimize component deliveries to the plant for the production of motor vehicles has been developed [18]. Reducing CO_2 emissions by 3% was simply achieved by lowering the number of lorries used for transportation and by increasing the lot sizes and the speed of the transportation and loading process. The impact of the type of forklift engine (diesel, gas, electric) on the nature of the impact on the environment was also assessed [19]. It was concluded that electric forklifts are more effective from an environmental point of view; however, the research did not take into account economic and technological factors (cost of forklifts, downtime for battery charging of forklift, etc.).

Analysis of the studies shows that a systemic approach to the solution of the problem is needed. This is especially true because once all of the complex subsystems of the production process and their interactions are taken into account, positive synergistic effect can be achieved.

3.3.2. *Optimization of technological processes on the assembly line*

Simulation models are used to determine the optimal parameters of technological processes when changing internal or external parameters of production. Input data for the development of the simulation model of technological process are typical manufacturing processes and Teamcenter database (data on assemblies, products, equipment, tools and environment).

The structure of individual technological process is adjusted in accordance with the composition and structure of the unified technological process by analyzing the need for each operation and the technological transition with the consistent refinement of all solutions. Technological design consists in the development of standard technological processes, from which in the future it is possible to assemble various methods of assembling cars. This makes it possible to significantly reduce labor input and the time required for their introduction into production.

Mass conveyor production is based on the principle of the flow organization of technological assembly processes, providing:

- the division of the assembly process into a series of assembly operations, sequentially arranged in time and space, performed by the operators-assemblers;
- the use of special transport devices to move the assembled units between assembly devices and to ensure a given assembly rate;
- the use of special transport devices for supplying parts and assemblies to the main assembly conveyor;
- the use of special and unified tools and devices for mechanization and automation of the technological process;
- mechanical machining of parts and assembly of units in machine-assembly shops.

When the production is organized in such way, assembly of the entire vehicle on the main assembly conveyor is carried out from the finished assembled units and aggregates, connected together by fasteners. The open architecture of the Teamcenter system allows you to connect to the PLM environment systems such as Matlab/Simulink and Rhapsody. In order to work with data in the usual formats, we can use the capabilities of dynamic integration with Microsoft Office software package. The obtained solutions are stored in the knowledge base and can be used in similar production situations.

The use of simulation models allows you to isolate operations that need optimization, determine the required number of employees and optimize the working load. Obtained solutions can allow reduction of the assembly time at the conveyor positions, and, accordingly, the cycle time by 6%, while the optimal loading of personnel decreases the number of errors.

Implementation of the proposed method was carried out during the development of the Decision Support System (DSS) for automotive company KAMAZ (of Naberezhnye Chelny, Russia) [20]. To optimize production processes, special documentation was developed. The documentation is integrated in production system for shared use [21].

Since the interdependent processes are modeled in parallel, program modules share the information for operational adjustment processes. Optimization of the conveyor in order to reduce delays is performed in two directions: alignment of operations on the assembly line positions and the operational management of the supply of components to the position. Optimization of technological transport includes providing the conveyor positions by necessary components with minimum cost (the number of forklifts and work time).

3.3.3. Monitoring and managing equipment efficiency

3.3.3.1. Literature review

One of the most important conditions of the successful operation for any industrial plant is to ensure uninterrupted operation of the equipment. That is why the indicators that characterize the quality of the equipment's use can be used as the objective function when modeling of technological processes [22]. In so doing, it should be taken into account not only the actual time of the equipment use and its performance but also the share of goods without any defects in the overall product output.

In the case of robotic production, the equipment efficiency depends significantly on the quality of its service, which affects indicators of availability, performance and quality of the final product. As a rule, a system to support the workability of equipment (maintenance and repair) is developed to ensure its efficient use. Frequency of service is determined depending on the equipment characteristics and is assigned by the manufacturer. To exclude catastrophic failure, the methods to predict and improve reliability exist.

Since there are different categories of losses, for monitoring the equipment condition, it is necessary to foresee methods for their control. The adjustment of the equipment maintenance system must be performed in accordance with the criteria of its efficiency. Furthermore, the method of complex multidimensional assessment of the performance indicators allows to raise the efficiency of production system management and, at the same time, to increase its stability as well as to reduce unplanned downtime.

The specificity of robotic production is that mistakes and failures in the technological processes are not attributable to mistakes of operators. Quality control of the equipment in this case can be carried out by comparison with the model of production system (i.e., system of virtual production). It is possible to identify the causes of errors by using imitation of the real system processes. In addition, simulation models allow to test new production concepts and agree with each other that all the subsystems on the design phase of production. It is also possible to optimize and to modernize virtually the existing complex of production with the aim, for example, to test the transition to the new product. Such systems allow to optimize the process of equipment maintenance, taking into account condition and features of the real system.

It is shown in the research paper [23] that maintenance scheduling, quality control and production scheduling influence one another and, therefore, need to be considered jointly for improving the system performance. A model has been created to integrate maintenance scheduling and process to develop a policy of decision management. It provided optimal parameters of preventive maintenance as well as a chart of control intervals, which minimize expected cost per time unit. Subsequently, the optimal interval of preventive maintenance is integrated with the production schedule in order to determine the optimal batch sequence, which allows to minimize penalty costs due to schedule delay.

The authors of the article [24] offer the method of multi-criterial classification of critical equipment (MCCE), which allows to classify equipment objectively according to its importance. They assume that according to such service approach the most critical equipment will not fail or, at least, all appearing failures will be rapidly detected and corrected in a minimum possible amount of time. To provide this information, the consequences of any failure in the appropriate equipment are analyzed for a particular company.

The article [25] is devoted to analyze mistakes and false operations, which require to carry out service operations. The authors designed two models to describe an ideal situation: the first one, in which a false-positive alarm implies the renewal of protection system and the second one: not. In the first situation, imperfect inspection is manifested in the scenario, where a false alarm implies an additional cost for the system owner; in the second situation, a false alarm does not imply renewal of the protection system. In both cases, a false-negative inspection can appear, in which the system is considered to be in a good condition, when it does not work in fact.

A condition-based preventive maintenance approach that is developed as a software service located in a "cloud" is presented in the paper [26]. It acquires and processes data from the shop-floor of machine tools using the technique of information fusion. The authors consider that benefits from the combination of monitoring and maintenance techniques under the umbrella of Cloud and mobile communication have not been still exploited sufficiently. At the same time, advanced maintenance methods, which cover and process the shop-floor information, can reduce costs and increase the sustainability of the enterprise. An operator reports through mobile devices the following data: status of machine tool (for example, available, busy, etc.), current running task, cutting-tool availability and appearance of failures. Combining inputs from the machine tool operator and the sensory system, an actual machining time of machine tools and cutting tools is calculated. The controlling data are processed through the technique of information fusion to identify the status of machine tool and, consequently, its actual machining time. On the basis of this information, the maintenance department is able to schedule the maintenance of machine tool according to its actual wear, not in fixed intervals.

As it is seen from the above review, the authors believe that to improve the efficiency and sustainability of the production system it is necessary to improve service. For these purposes, we offer different methods, but the general is the simulation of systems. A model is suggested to be constructed using statistics of failures and malfunctions of equipment. Reliable operation of the equipment is based on information about its failures, unplanned stops and failures recorded in the complaint acts and stored in databases. The analysis of such information allows us to identify the causes of emergencies and warn them. For the operational management and adjustment of a service system, intelligent control systems, including cloud technology, are offered.

3.3.3.2. Challenges in transition to a smart factory

In our opinion, the existing platforms to create a unified information space are the most effective way of both strategic and operational management stages of the life cycle of the product. The concentration of heterogeneous data, semantic linking and providing access to them through search interfaces is a topic which today is engaged in many companies. Many long-term benefits of implementing management systems and product lifecycle (PLM) cannot be achieved without having a comprehensive digital manufacturing strategy, which allows the simulation of production processes aimed at reuse of existing knowledge and optimizes processes before products are manufactured. In addition, digital production allows you to get feedback from actual manufacturing operations and to incorporate it into the design process of the product, so businesses already can solve production and technological problems at the design stage. Among the initiatives for the development of systems to support digital manufacturing is improving user interaction by providing information in the context of the problem being solved, thanks to which engineers can make the right decisions faster. Measures are taken to ensure direct communication with technological equipment, such as programmable logic controllers (PLCs), machine controllers, numerical control (CNC), etc. Single platforms are created to manage the information stored in the PLM system and manufacturing execution systems (Manufacturing Execution systems, MES).

The creation of DSS is especially important when creating flexible production systems, which are based on robotic systems. However, it is necessary to consider that the flexible automated manufacturing (FAM) operates on the basis of a solitary technology, so the work of all production components is coordinated as a whole multi-level control system which ensure the change in program, fast tuning technologies when changing production facilities.

The decision on the design and optimization of manufacturing from Tecnomatix is parametric 3D smart objects that can be used for quick and efficient planning of the enterprise. The use of 3D smart objects for planning of the enterprise also enables to detect design errors not on the factory floor but even in the planning stage. The flow of materials, transportation, logistics and auxiliary operation can be optimized using material flow analysis and modeling events.

Along with it, it should be borne in mind that while the transition to a smart factory, there will be a number of problems that need to be solved.

The product/process design is the first of these challenges. It covers all the tools and engineering services supporting the design of parts, finished products, processes, production lines and factories. This category is evolving from a separate design towards one that is largely modeled and simulated, and most importantly, joint between product and process.

The virtual factory means simulating the production lines from start to end and enabling you to anticipate potential sources of additional costs or poor quality and preset the machinery and equipment and control of the productive facilities. This integrates the control system (digital control, interconnection with the factory's upstream and downstream, planning and centralized control of the production line), traceability (sensors of production conditions, monitoring of individual parts being manufactured) and the management of physical flows (automation of internal logistics and interconnection of external logistics).

Manufacturing operations (at the heart of processing) were identified as the third challenge, with two performance criteria: precision (the optimization of existing technologies, such as high-speed machining and laser cutting, and smart self-correcting machines) and flexibility (multi-device multimedia additive manufacturing machines).

The fourth challenge covers services related to the productive facilities. It includes the integration services of the various components of the production line and the installation and maintenance of production machinery.

The fifth challenge is the newcomer; it includes the digital technologies behind the upheaval. Hard to imagine the factory of tomorrow without the contribution of cloud computing, whether to store data, work with remote desktops or use SaaS software; without Big Data Analytics, which will improve production through predictive remote maintenance or will increase energy efficiency and without the Manufacturing Internet of Things, these autonomous cyber-objects capable of making local decisions.

Finally, work organization is the last challenge. Examples include the establishment of organizations that empower operators or can learn.

3.3.3.3. *Examples of existing cyber-physical systems*

When developing a strategy for transition to a smart factory, it should be borne in mind that modern industrial automation systems are composed of several clearly separated levels:

- Data collection level (sensors and actuators)
- Level of control (operator terminals and control devices)
- Level of business processes management (computers for data processing, MES systems)
- Manufacturing level (server, where MRP I, II, III, ERP systems, etc. are located).

Each of these levels is relatively well structured and individual devices can be clearly mapped to one of the levels.

In Industry 4.0, the system structure changes. The data collection level remains a separate dedicated level, as it is now, but the devices will be more intelligent and they will also significantly increase in numbers. All other functions will move to the high speed real-time network consisting of data processing center and cloud computing. The benefits of such a structure are as follows: (1) reduction of diversity of devices and processing hardware that are the most modern in the world; (2) separation of specific functions and (3) the use of augmented and virtual reality. All of it contribute to simplification of the management process, more efficient use of resources and, consequently, cost savings. This approach has not been implemented yet due to the low efficiency, reliability and throughput of communication channels between servers and data collecting devices. However, all these problems will be solved in new and future systems.

Industry 4.0 will be built in cyber-physical systems, which involves the integration of computation, networking and physical processes. Embedded computers and networks monitor and control the physical processes, with feedback loops where physical processes affect computations and vice versa.

An example of such a system today is the CarTel project at MIT [27] where a fleet of taxis collects real-time traffic information in the Boston area. This information is combined with historical data to calculate the fastest routes for particular times of the day. Another example that you may be familiar with is the Smart Grid. One of its definitions, based on [28], is: "A modernized electrical grid that uses information and communications technology to gather and act on information in an automated fashion … to improve the efficiency, reliability, economics, and sustainability of the production and distribution of electricity."

Finally, an example for a factory [29] is changing systems so that the energy consumption in a vehicle assembly line is reduced when the line does not operate. Today, many production lines continue running during breaks and weekends. Consider laser welding technology that remains powered up over weekends, so it can resume quickly on Monday. This practice consumes up to 12% of total energy consumption of the assembly line. With Industry 4.0 and cyber-physical systems, robots will go into standby mode as a matter of course during short production breaks and power down during longer breaks. Speed-controlled motors that reduce the energy required to run machines will be widespread. Such changes will significantly reduce energy consumption and will be taken into account up front as part of Smart Factory design practices.

4. Conclusions

Introduction of the concept Industry 4.0 can reduce losses and improve efficiency of the processes, because only comprehensive solutions can lead to real improvements. Developers of smart vehicles optimistically believe that autonomous vehicles on roads will improve the safety of the transport system, reducing the number of accidents due to the exclusion of the human factor.

At the same time, despite the existing positive experience of introducing intelligent technologies, there are still can appear situations when alleged improvements can lead to losses while vehicles production. There are also a lot of issues identified by analysts that could lead to critical situations while such vehicles operation.

First, it should be noted that operation of any complicated system is always closely connected to the risks. It is especially actual for transport systems. The complexity of transportation systems' risk analysis is due to the fact that an accident potentially may happen in any part of the route and the same events may lead to absolutely different consequences. That is why every decision for the existing transportation system's optimization should also be considered from the perspectives of risk management.

The most possible risks with the most drastic consequences can be grouped by types:

1. Technical:
   a. Reduction of operational reliability because of increased complexity of the vehicles' design.
   b. Increased infrastructure requirements.
   c. Increased requirements to communication systems.
2. Ecological:
   a. The risk of technological disasters if there are cyberattacks or failures in the control system.
   b. The risk of increasing negative impact on environment because of expansion of the vehicles' fleet.
3. Organizational:
   a. Complexity of the movement algorithms for rough terrain.
   b. The absence of a panoramic view of the streets, impeding the routing.
   c. Increased requirements to information processing speed.
   d. Complexity of decision-making in unusual situations.
4. Economic:
   a. The high cost of infrastructure changes.
   b. The high price of the vehicles.
5. Legal and ethical:
   a. Ambiguity of legal responsibility for causing damage and when organizing transportation.
   b. Loss of privacy.

6. Social:

   a. The loss of jobs by people whose work is related to driving vehicles.

   b. Lack of driving experience of drivers in critical situations.

   c. Loss of self-driving capability.

All problems described above have to be solved now, before the widespread of intelligent vehicles. The main but not the only trends in solving these problems are as follows:

- Development of new communication systems, information security systems and improving the reliability of control systems.
- The use of resource-saving technologies, improving the efficiency of transportation management.
- Accidents' statistics collection and analysis, development of expert systems and knowledge bases in the field of traffic management and improving the movement algorithms and the vehicles' design and position control systems.
- Service system improvement, the use of highly reliable components and redundancy of elements that ensure safety.
- Improvement of gesture recognition and speech-understanding technologies.

In this regard, the processes of the vehicles' intellectualization should be considered from the point of view of each stage of the life cycle and the most dangerous situations should be highlighted for the subsequent countermeasures development.

Author details

Irina Makarova, Ksenia Shubenkova*, Vadim Mavrin and Eduard Mukhametdinov

*Address all correspondence to: ksenia.shubenkova@gmail.com

Kazan Federal University, Naberezhnye Chelny, Russia

References

[1] Department of Economic and Social Affairs. World Population Prospects the 2015 Revision: Key Findings and Advance Tables. New York: United Nations; 2015 66 p

[2] City Population. Major Agglomerations of the World [Internet]. 01-01-2017. Available from: https://www.citypopulation.de/world/Agglomerations.html [Accessed: 24-07-2017]

[3] UNEP. Summary of the Sixth Global Environment Outlook GEO-6. Regional Assessments: Key Findings and Policy Messages [Internet]. 2016. Available from: http://www.unep.org/geo/assessments/regional [Accessed: 24-07-2017]

[4] UNEP. More than Half of World's Population Now Living in Urban Areas, UN Survey Finds [Internet]. 2014. Available from: http://www.un.org/apps/news/story.asp?NewsID=48240#.WXbkYRXyjMw [Accessed: 24-07-2017]

[5] Makarova I, Pashkevich A, Shubenkova K. Ensuring sustainability of public transport system through rational management. Procedia Engineering. 2016;**178**:137-146

[6] Makarova I, Khabibullin R, Pashkevich A, et al. Modeling as a method to improve road safety during mass events. Transportation Research Procedia. 2016;**20**:430-435

[7] Makarova I, Shubenkova K, Mavrin V, et al. Ways to increase sustainability of the transportation system. Journal of Applied Engineering Science. 2017;**15**:89-98

[8] Makarova I, Shubenkova K, Gabsalikhova L. Analysis of the city transport system's development strategy design principles with account of risks and specific features of spatial development. Transport Problems. 2017;**12**(1):125-138

[9] Makarova I, Khabibullin R, Shubenkova K, et al. Ensuring sustainability of the city transportation system: Problems and solutions (ICSC). E3S Web of Conferences. 2016;**6**:02004

[10] Makarova I, Khabibullin R, Belyaev E, et al. Increase of City transport system management efficiency with application of modeling methods and data intellectual analysis. In: Sladkowski A, Pamula W, editors. Intelligent Transportation Systems—Problems and Perspectives. Switzerland: Springer; 2016. p. 37-80

[11] Richardson N et al. Assessing truck drivers' and fleet managers' opinions towards highly automated driving. Advances in Human Aspects of Transportation. 2016;**484**:473-484

[12] TechNet. Explanatory Note to the Road Map of the National Technological Initiative [Internet]. 2015. Available from: http://assets.fea.ru/uploads/fea/nti/docs/2015_1225_Zapiska_technet.pdf [Accessed: 24-07-2017]

[13] Paavel M, Snatkin A, Karjust K. PLM optimization with cooperation of PMS in production stage. Archives of Materials Science and Engineering. 2013;**60**(1):38-45

[14] Li D, Zhang C, Shao X, Lin W. A multi-objective TLBO algorithm for balancing two-sided assembly line with multiple constraints. Journal of Intelligent Manufacturing. 2016;**27**(4):725-739

[15] Amen M. Heuristic methods for cost-oriented assembly line balancing: A comparison on solution quality and computing time. International Journal of Production Economics. 2001;**69**(3):255-264

[16] Padrón M, Irizarry M, Resto P, Mejía H. A methodology for cost-oriented assembly line balancing problems. Journal of Manufacturing Technology Management. 2009;**20**(8):1147-1165

[17] Erel E, Sabuncuoglu I, Sekerci H. Stochastic assembly line balancing using beam search. International Journal of Production Research. 2005;**43**(7):1411-1426

[18] Florian M, Kemper J, Sihn W, Hellingrath B. Concept of transport-oriented scheduling for reduction of inbound logistics traffic in the automotive industries. CIRP Journal of Manufacturing Science and Technology. 2011;**4**(3):252-257

[19] Fuc P, Kurczewski P, Lewandowska A, et al. An environmental life cycle assessment of forklift operation: A well-to-wheel analysis. The International Journal of Life Cycle Assessment. 2016;**21**(4):1-14

[20] Kamaz PTC. Official Website [Internet]. 1997-2017. Available from: https://kamaz.ru/en/ [Accessed: 24-07-2017]

[21] Khabibullin R, Makarova I, Pashkevich A, et al. Application of simulation modeling to improve management of technological processes during production of automotive components. In: Maga D, Stefek A, Brezina T, editors. Proceedings of the 2016 17th International Conference on Mechatronics Mechatronika (ME); Dec 07-09, 2016; Prague, Czech Republic. 2016. p. 43-49

[22] Makarova I, Khabibullin R, Mukhametdinov E, et al. Efficiency management of robotic production processes at automotive industry. In: Maga D, Stefek A, Brezina T, editors. Proceedings of the 2016 17th International Conference on Mechatronics Mechatronika (ME); Dec 07-09, 2016; Prague, Czech Republic. 2016. p. 35-42

[23] Pandey D, Kulkarni M, Vrat P. A methodology for joint optimization for maintenance planning, process quality and production scheduling. Computers & Industrial Engineering. 2011;**61**(4):1098-1106

[24] de Leon FG, Cartagena J. Maintenance strategy based on a multicriterion classification of equipments. Reliability Engineering & System Safety. 2006;**91**(4):444-451

[25] Berrade M, Cavalcante C, Scarf P. Maintenance scheduling of a protection system subject to imperfect inspection and replacement. European Journal of Operational Research. 2012;**218**(3):716-725

[26] Mourtzis D, Vlachou E, et al. A cloud-based approach for maintenance of machine tools and equipment based on shop-floor monitoring. Procedia CIRP. 2016;**41**:655-660

[27] MIT Cartel. CarTel [Internet]. [Updated: 2014]. Available from: http://cartel.csail.mit.edu/doku.php [Accessed: 24-07-2017]

[28] Department of Energy. Smart Grid [Internet]. Available from: https://energy.gov/science-innovation/electric-power/smart-grid [Accessed: 24-07-2017]

[29] KennisBank. Best Practices [Internet]. Available from: http://www.3if.be/nl/kennisbank

Section 4

The Influence of Social and Economic Factors in Urban Space Conception

Chapter 8

Life Cycle Insights for Creating Sustainable Cities

Getachew Assefa

Additional information is available at the end of the chapter

http://dx.doi.org/10.5772/intechopen.81633

Abstract

The chapter focuses on the full life cycle of material and energy flows and uses in cities. Most of the impacts and opportunities in making cities more sustainable exist in how and what types of materials and energy source are used. The life cycle perspective of materials used in buildings and infrastructure systems is better addressed at the point of planning and design. The energy aspect touches both the efficiency of utilization and the impact intensity of the energy used to power and heat urban spaces and fuel transport systems. The type of sources of upstream supply of materials and energy is thus crucial. The commendable efficiency measures targeting the operation phase of urban systems should be accompanied by a consideration of the embodied impact of materials used and the end-of-life management of the materials following the end of their service life. The chapter ends with recommendations on best practices that potentially leverage on life cycle assessment results. It also covers the merits of employing the social life cycle perspective together with the environmental life cycle and economic life cycle in a life cycle sustainability assessment framework that seeks to define the triple bottom line space of lower unsustainability conditions.

Keywords: life cycle assessment, life cycle sustainability assessment, climate change, greenhouse gases, circular economy, sustainable cities, material flows

1. Introduction

Addressing increasingly pressing global challenges of climate change and other long-term impacts demands critical understanding of the sustainability (environmental, social, and economic) performance of current best practices of developing and managing the totality of urban infrastructure systems and accounting for how future technical and nontechnical changes influence the sustainability of cities. Dominant practices currently focus only on one aspect of the sustainability (e.g., environmental [1]) or only one component of the city at a time (e.g., transportation [2]) or only part of the life cycle of the infrastructure (e.g., embodied energy [3]). Decisions based on such partial information and limited knowledge should be replaced by more complete knowledge as a basis for decision analysis of the development and management of future infrastructure systems in cities to curb damaging and costly suboptimal developments. A comprehensive account of relevant environmental, social, and economic aspects of our urban systems provides municipalities and high levels of government with the opportunity to look at the merits, synergies, and trade-offs associated with alternative development and operational decisions in creating sustainable cities.

What does a sustainable city look like? The literature offers as little on a widely accepted definition of sustainable city as it does on what sustainable development in general is. The difficulty of defining sustainable development has led to over 200 definitions [4].

Closer examination of the literature over the past 20 plus years shows that different issues are emphasized in the urban sustainability discourse depending on field of study and the problem at hand. Land-use system as a vehicle for achieving urban sustainability was not widely supported with evidence as mentioned in work published in 1997 [5]. When areas where improved environmental performance is needed in cities were identified, the emphasis was on how to integrate them with social, economic, and political goals of sustainable development [6]. In a case study on the Scottish capital of Edinburgh [7], a work on weak and strong sustainability measures for cities differentiated between unsustainable cities, weak sustainable cities, and strong sustainable cities. The strong sustainable cities model is close to what is now promoted as urban scale equivalent of circular economy at the majority of materials reused, repaired, and recycled, recognizing the limits to resources and assimilation capacities (e.g., see [8]).

The literature around sustainable cities reflects the compact city concepts of abating urban sprawl, increasing density, and achieving mixed uses while ensuring diversity and vitality, both socially and economically [9]. Urban form is an important element that has attracted significant research attention in the literature. Policies that focus on sustainability of cities should stay away from the quest for a long-term goal of creating a single definitive urban form that is sustainable as it does not exist [9]. There are more avenues that lead to more sustainable forms. In their research that involved many British cities on the role of urban form, their conclusion was that "there are no simple answers or clear relationships between urban form and the dimensions of the sustainable city." Others have also explored how density and other urban forms relate to sustainability (e.g., [10]).

Two important aspects of urban sustainability are the issue of system boundary and the issue scale. Allocation of greenhouse gases (GHGs) from surface and air transportation spatially to cities that make up a commuter shed is demonstrated in [11]. Factors such as surface vehicle miles traveled per capita (estimated to range from 13 to 129 km/capita/day for the Denver commuter shed) that affect the allocation are identified. Along the same line, geography, urban form, and the urban economy affect the GHG emissions of cities were highlighted elsewhere [12].

While liveability of cities is seen within a broader framework of urban sustainable development [13], the role of sufficiently incorporated urban natural areas in creating sustainable cities is also stressed [14]. Sustainable cities are also presented in different studies as low-carbon cities (e.g., [15]). As part of achieving sustainable development of Chinese cities, low carbon goals of urban development are proposed by local governments [16]. Governance level is considered a critical element in other parts of the world as well [17–19]. They provide the examples of "garden city, Chinese electric valley, and Sun City." Examining how the concepts of compact city, the zero-carbon eco-city, and the U-ecocity contradict or fall short of meeting sustainability principles, the use of the term "transition toward the sustainable city" was proposed as an accurate and more effective goal of our cities around the world aspire to be [20]. Arguments for establishing guiding principles at different levels in creating sustainable cities are abundant (e.g. [21, 22]).

The literature is full of connecting the concept of sustainable cities to the triple bottom line of social, economic, and environmental standing of different aspects of urban areas. The need to convert our unsustainable cities by developing sustainable city models and rehabilitation plans that help us implement sustainability practices considering environmental and social performance factoring in economic elements is pointed out [23]. Earlier, the social, economic, and environmental dimensions of urban sustainability feature both complementarity and contradiction, underlining the importance of investigating alternatives using a holistic approach [9]. Even much earlier, going beyond the triple of bottom line of social, economic, and environmental sustainability, the n-bottom line framework of performance attributes used in the research, development, and demonstration activities under the sustainable cities research theme at CSIRO, Australia, was presented [24].

Overall, there is a recognition that cities are better supported with improved sustainability-oriented policy-relevant information and knowledge on urban design, development, and management (e.g., [25]). There are potential multiple features that can represent what can be considered as part of the sustainable city vision [26].

From a life cycle perspective, a sustainable city is here defined as one whose residents and decision-makers account for a life cycle triple bottom line (social, economic, and environmental bottom lines) performance in prioritization and ranking of alternative decisions emphasizing the process. A sustainable city as a product is one that advances socially and economically with a nondeclining environmental quality over time. Environmental sustainability that embodies environmental quality, or rather the avoidance of the degradation of it, is measured in the form of environmental impacts such as climate change, acidification, eutrophication, fossil fuel energy use, resource depletion, smog, ozone layer depletion, and different forms of toxicity. The level of impact should be capped within the city's allocated share based on what is required for achieving global net minimum sustainability threshold. Economic progress as a proxy to economic sustainability in a city in this case is measured as increase in GDP per capita together with very low or decrease in GINI coefficient implies. The economic progress should be at the level where the city continues to make an operational profit that enables it to sustain its activities indefinitely. Social sustainability is represented by different aspects of social well-being such as inclusiveness, participation, employment opportunities, education successes, and health service coverage. It can be monitored by higher level of measured resident's satisfaction with sustaining their social well-being at the time of measurement and in the long run.

The rest of the chapter continues with a brief outline of challenges in cities, presentation of some life cycle opportunities, discussion of areas to be considered in achieving a life cycle sustainability of cities, and ending with the way forward.

2. Urban challenges

Cities operate in a resource-constrained platform that is full of challenges as they work with land use and planning priorities. Among the different ways to structure the challenges of cities, one conventional approach is to categorize issues by type of city department responsible for addressing them. Thermodynamically, challenges in our cities can be linked to the quantity and quality of material and energy generated and/or consumed in cities. These material and energy flows affect land use in and beyond the city boundaries. Creating a sustainable city is ensuring how it deals with its material and energy flows and how its physical assets are developed and managed from a life cycle perspective. It is about realizing ways of developing and managing sustainable buildings, water supply, waste management, wastewater treatment, and transport systems. It is about how a city serves its residents and how its economy is managed. The challenges are presented here in terms of material and energy flows entering, occurring, and leaving cities.

2.1. Material aspects of cities

The material dimension of cities includes the buildings, the infrastructure, green spaces and parks, and the transport. Under infrastructure, roads, city rails, water infrastructure, and wastewater infrastructure are covered. It covers both bound materials and material flows through cities. From concrete to steel, huge amounts of materials reside in and pass through mega cities [27]. The amount and type of materials in the building and infrastructures has been accumulating without meaningful thought concerning their fate. Traditionally, there were no regulatory requirements or economic or other forms of incentives to force or encourage material specifiers, builders, etc. to account for the end of life of the construction materials after the service life of the building or infrastructure is over. The toxicity buildup in these materials is worrisome, should the emitted substances make their way to our surface and groundwater or to the atmosphere.

When human beings used materials to build in the early days of urbanization in Asia and Africa, the materials were dominantly monotype and were loosely connected. Biomass in the form of straw, wooden beams and lintels, sun-dried clay mud, and stones consisted of most of these building materials. The materials, if necessary, could be disconnected easily and reused more than once. These materials were also harvested and collected from the immediate environment locally. With the advent of synthesis chemistry and the introduction of composite materials, the complexity of construction materials has increased significantly. The subsequent construction technologies of meshing up different types of materials have led to inseparable end products that can no longer be reused in a second life. The distances from which materials are sourced have also increased significantly for most of the supplies.

What we do in cities today will be critical in terms of influencing the material content of the existing building and infrastructure stock, which will stay for the next 50 years or more depending on the life time of the building and infrastructure. In the past and still in most cases, majority of these materials after demolition are sent to landfills. For lack of space and not-in-my-backyard pushes from local communities, cities are finding it difficult to get landfilling spaces to cope with the increasing quantity of building and demolition waste.

Regulatory restrictions and increasing awareness will force municipal decision-makers and developers to find new ways of sourcing new materials and using them in new constructions. Construction technologies need to account for the eventual disassembly of materials used in buildings and infrastructures and separation by material type.

Our knowledge of what materials contain and how they should be connected and used in buildings and infrastructure systems should account for how they will be disposed of at a later stage. Cities can save on extraction of virgin raw material and avoid other resources consumed during the extraction, processing, and transport of materials.

2.2. Water and wastewater aspects of cities

Many urban areas around the world do not provide basic services of clean water supply and adequate sanitation. In 2015, 2.1 billion people lacked access to safely managed drinking water [28]. Waterborne diseases are responsible for a significant number of deaths and lost productivity and shortened life expectancy in many developing countries. Almost 1000 days die of waterborne diseases associated with lack of appropriate sanitation and clean water before celebrating their fifth birthday [28].

Cities have been from the very beginning on a wasteful trajectory when it comes to water supply and treatment. One aspect of this resource leakage is that our urban areas continue to utilize potable water treated through a series of technologies and consuming energy and chemicals for treatment only to be used to flush toilets. We are so locked-in to this obsolete system that our plumbing education and building codes are tied to this unduly common and old inefficient practice. The dependence on these paths and system has prevented decentralized innovative solutions from making important contributions both in incremental and transformative changes of how water and wastewater services are delivered in cities. The solutions that are proposed to areas where there are poor and no such services at all are often large and centralized systems. Due to increased urbanization, many cities are overstretched in terms of water and wastewater treatment plant capacity. Building new systems and upgrading these large systems require big investments. The amount of energy consumed in the treatment of water and wastewater and the transport and distribution is one hindrance in operating and maintaining existing systems in cities where topography is challenging. Nutrient leakage associated with wastewater management is another aspect that needs attention.

2.3. Solid waste aspects of cities

The solid waste problem has economic implications in many developed cities. In developing countries, the issue takes a multifaceted form as it affects the social, environmental, and short-term and long-term economic advancements of urban areas. With increasing urbanization, the composition and quantity of waste is surpassing the already meager resource allocated to proper management of the waste in many cities of the developing world. According to the UN, the share of the world's population living in cities is expected to rise from 54 percent in 2014 to around 66 percent in 2050 [29].

The rate of increase of volume of solid waste generation is higher than the rate of urbanization as established by a World Bank report from 2012 [30]. By 2025, the planet will have 4.3 billion urban residents generating about 1.42 kg/capita/day of municipal solid waste, that is, 2.2 billion tons per year. This increase in solid waste, which is the single largest budget item in many municipalities, will create unprecedented stress in cities in many developing countries, which are already operating under capacity in properly managing their solid waste. If current trends continue, a badly needed global "peak waste" will not happen this century [31].

2.4. Energy aspects of city buildings and transport

The heating, electricity, and fuel consumption of cities dominates the overall consumption of energy by society. Cities in the cold climate zones of the world specifically account for higher consumption of heat. The hottest regions of the world consume significantly high amounts of ventilation and air-conditioning energy. Decisions about the type and quantity of fuels used to deliver the relevant energy services often revolve around their energy intensity and impact intensity. Cities in different countries therefore have legislated to limit the amount of energy consumed by buildings. There has also been an increasing tendency of moving away from fossil fuel consumption in heating and electricity generation. Those in the cold climate regions will have to plan how their buildings and infrastructure will be heated during the long winter months of the year. The long-life time of buildings and the associated lock-in effect of the existing building stock require innovative approaches to redevelopment and urban renewable plans.

As the power grid decarbonizes, many cities in the world will continue to deal with their transport fuel. The amount of fossil fuel consumed for transportation in cities across the world is still the dominant contributor in both energy consumption and associated emissions of different types. This can be attributed to the high-carbon fuels and the increased mobility and expanding urbanization-led infrastructure systems. The expansion of these systems contributes to increase in indirect consumptions of energy through increased economic activities triggered by the infrastructures.

Measures aiming at energy efficiencies may not necessarily lead to the desired outcome of net reduction of energy and associated impact at the society level because of a potential rebound effect. Any extra money associated with energy efficiency when spent on previously nonexistent activity comes with higher energy consumption such as long-distance leisure travel; any gain from the energy efficiency is offset by the new energy-intensive activity and undermines the otherwise positive energy efficiency program.

Legislative and incentive instruments aiming at reducing energy use in cities with focus on buildings and transport systems better take a bigger picture and look beyond the different subsystems and factors including technological, architectural, urban form, lifestyle, behavioral, and cultural aspects (e.g., [32]).

3. Life cycle-based solutions for sustainable cities

Many of the problems that are established as critical in the context of urban areas are characterized as severe based on partial considerations of the full life cycle. They can be worse when accounted in their totality from a life cycle systems' perspective considering the downstream and upstream systems as well. Cities at the same time have a capacity of leverage in moving society at large in the right direction. Dematerialization of the urban infrastructure, exploring the area of integrated infrastructure and life cycle-based performance labeling are discussed as important elements of such opportunities. Capitalizing on such opportunities, however, demands a paradigm shift in the way cities do business.

3.1. Dematerialization of urban infrastructure

Innovative designs can be sought to reduce the quantity of materials extracted, processed, and utilized without compromising the quality of the infrastructure. This reduction in the amount of materials is not limited to what is finally bounded in the urban form. A life cycle perspective provides the opportunity to take stock of the materials that are wasted upstream in the mining and quarry sites as well as the waste that should have been diverted from landfills in and around cities. The life cycle lens goes beyond the embodied material during the preuse construction phase. Services provided by and on urban infrastructure should also be dematerialized. Dematerialization is better seen as covering both a relative and absolute decoupling of resource consumption and associated environmental impact from economic growth. It is realized through concerted efforts of achieving significant increase in material and environmental efficiency. At the broadest level, urban areas around the world should be developed and operated with an additional type of decoupling in mind that decouples human well-being from economic growth and consumption through a set of measures that include reduction of excessive consumption levels. Decarbonization as a specific case of dematerialization is best applied in the form of decarbonizing the energy and transport systems, which are the top two contributors for greenhouse gases in cities around the world.

3.2. Assessment integration and integrated infrastructure

Given existing building and infrastructure stock are here to stay for long, innovative technical and financial mechanisms are required to reuse old infrastructure systems and as last resort recycle materials from them. Repurposing should be a prioritized norm in renewal and redevelopment works of urban centers across the world. The task of repurposing will be increasingly huge with game-changing technical transformation globally. One good example is the large-scale introduction of autonomous vehicles. The need for our massive parking surface and above surface structures will be reduced significantly, and large part of these structures will be abandoned. It is, thus, time to rethink new purposes for them and redesign and redevelop them for a second life and beyond. New projects can be designed and implemented using advanced knowledge using our experience and by finding synergies between different urban forms. Best practices around the world should be emulated with focused attention to local contexts and variables. Urban decision-makers will benefit from twofold integration in relation to infrastructure systems in cities: integration of assessment aspects and integration potentials of infrastructure systems.

Under the assessment integration, elaborating the concept of a comprehensive integrated assessment in relation to decision analysis of infrastructure development and management is important. Setting higher environmental, social, and economic standard on our cities requires the integration of decision-making related to urban infrastructure systems by looking at the triple bottom line as a more comprehensive platform for complementary consideration of the most important and relevant variables while avoiding double counting of the aspects informed by such variables. Understanding the sustainability (environmental, social, and economic) performance of current best practices of developing and managing the totality of urban infrastructure systems and how future technical and nontechnical changes influence the sustainability is crucial. Conventional practices focus either on only one aspect of the sustainability or on only one infrastructure at a time and/or only part of the life cycle of the infrastructure. Decisions based on such partial information and circumscribed knowledge should be replaced by more complete knowledge as an input to the decision analysis stage of the development and management of future infrastructure systems. A comprehensive account of relevant environmental, social, and economic aspects of our urban infrastructure provides municipalities and higher level of governments the opportunity to look at the opportunities, the synergies, and trade-offs associated with alternative development and operational decisions. Our understanding of the performance of the current best practices of developing and managing infrastructure systems from environmental, social, and economic perspective is weak. Even weaker is the level of knowledge we have around the changes in performance in the future with the introduction of technical and nontechnical shifts that can potentially happen in the next 10–30 years. Without an actionable knowledge regarding the current and future potential performance of our cities, we will fall short of getting it right in terms of what measures will be critical in containing the negative impacts of disruptions that affect infrastructure systems from a life cycle perspective. Integrated assessment provides the opportunity of identifying cobenefits and adverse side effects, which will otherwise lie outside conventional assessment lens (see, e.g., [32]).

On integrated infrastructure, exploring how different levels and scales of physical integration, data integration, resource integration, management integration, and other forms of integration affects the overall performance of infrastructure systems is critical. In some cities, there is already a demonstration of integration of infrastructure systems through connecting material and energy flows of individual systems [33]. Connecting the waste management infrastructure with the transport system using fuels such as biogas produced from waste to power vehicles is one example. Or the use of waste-driven electricity to power transport systems could link the three infrastructure systems through material and energy flows. Advanced integration will be physical (e.g., surface) integration. The future which in many areas is already here holds promising technologies in terms of integrated infrastructure such as solar shingles and solar panel roads. The photovoltaic layered roads or pedestrian ways, for example, can increase the values of infrastructure by adding new functions or layering functions, which is the basis of life cycle assessment's focus on functions of product systems. Envisioning new vehicle technologies that can charge while driving wireless/wire free from the roads and even from other autonomously driven vehicles is not wild. These kinds of developments will, of course, require all kinds of new fiscal, regulatory, and social transformation to work effectively.

3.3. Life cycle-based labeling and certification

As cities and nations increasingly set greenhouse gases-focused goals and broader sustainability targets, there is a need for measuring, monitoring, and communicating performances and progresses made. Increased sustainability awareness of citizens and the demand for better options offered by cities and parts of cities to prospective residents would encourage the development of sustainability rated or certified infrastructures, districts, neighborhoods, and cities in the future. Assessments that can potentially support such ratings and certifications come in different shapes and sizes depending on the purpose, scope, and object of assessment. Life cycle assessment is about evaluating the environmental impacts of product systems such as buildings and infrastructure systems over their life cycle. Environmental impacts covered in a comprehensive life cycle assessment include climate change, ozone layer depletion, smog, eutrophication, acidification, human toxicity, ecotoxicity, biotic resource depletion, abiotic resource depletion, and fossil fuel depletion. Life cycle sustainability builds on life cycle assessment to include life cycle costing and social life cycle assessment.

Metrics for monitoring and communicating supported by quantitative data are helpful. There is, however, a need for scrutinizing the quality of data, for example, by qualifying the assessment results and conclusions. More and better data collection from primary sources will be useful in addressing the kind of uncertainty associated with, for example, urban greenhouse gases inventory [34].

Verifiable and independently reviewed sustainability labeling and certification of buildings, neighborhoods, infrastructure systems, subcities, and cities has the potential of sending the right signals to the market and thereby fostering the expansion of best practices and state-of-the-art design, development, and management of the different levels of organization of our urban areas. The performance labeling of infrastructure systems and other elements of urban forms will benefit from the experience of working with the life cycle-based environmental product declarations (EPDs) in the building sector. Rules that set the principles and requirements to be followed in developing EPDs are set in product category rules (PCRs) established through deliberative and participatory processes involving different stakeholders including relevant industrial associations. Both PCRs and EPDs as a basis for verification, comparison, and evidence-based monitoring are developed as living documents and are updated from time to time to capture new data and knowledge, new technology, and new requirements.

4. Toward life cycle sustainability of cities

Urban sustainability is both about how to make new developments sustainable while renewing existing building and infrastructure stock to fit the sustainability requirements of future cities. Urban areas of the future will be dominated by self-driving vehicles, renewable powered homes, and artificial intelligence-enabled systems and service. They will be better defined by how much of their physical assets and material flows in cities are in line with the principles of circular economy. Cities are well positioned to catalyze the institutional and technical aspects of rolling out a well-thought strategy for realizing an urban circular economy that covers physical assets in the city, products and services that are produced and/or consumed or pass through it. Circular economy is about ensuring a full account of the design, production, distribution, use, reuse/repair, and recycling with few net inputs of raw materials. Many cities around the world are at the bottom of the pyramid of hierarchy of material efficiency broadly known as "reduce, reuse, recycle, recover, and dispose," which is at the core of the circular economy as materials and consumables are still disposed in open dumps in many developing countries.

Three areas of interest will drive the path to the development of more sustainable cities: triple bottom line space of better sustainability conditions; life cycle data access and quality; and streamlined semiquantitative life cycle evaluations.

4.1. Triple bottom line space of better sustainability conditions

Decision-makers at different levels struggle on regular basis with multitude of trade-offs. They are better supported by tools and frameworks that show beyond impact assessment results. A decision-maker wants to know as to what makes a given alternative better or the best of the pool and what elements should be in place to make the selected alternative work. For new technologies, designs, and development proposal, the need for such appraisal is more critical.

The concept of triple bottom line space of better sustainability conditions is here introduced as a suit of social, economic, and environmental conditions that a development proposal or a redevelopment plan of an existing part of a city should meet to function based on life cycle sustainability principles. The concept refers to a performance space delimited by the boundaries and thresholds of environmental, economic, and social conditions that can potentially encourage implementation of new ideas, new technologies, and novel solutions in urban areas around the world. These boundaries can be set based on context-specific analysis that leads to city-wide net sustainability. The premise is that urban development options and pathways with triple bottom line performance numbers within the boundary conditions have higher chance of broader management and public buy-ins than those that lie outside the space boundary conditions. Research is required on how to establish the lower and upper boundaries of space in view of creating the triple bottom line feasibility considering policy, market forces, and consumer perspectives.

Under social conditions are the sustainability-relevant behaviors of residents and barriers associated with product-related and lifestyle and culture-oriented practices. These include behaviors that lead to decrease or increase in energy and water consumption, recycling and composting waste, and supporting wildlife in gardens; travel behavior and car ownership; social participation; and the use of local services, businesses, and facilities [9]. Under economic conditions are city activities that lead to a per capita income level that allows a sustainable level of consumption and a rate of job creation that agrees with the rate of increase in labor force. The environmental conditions allude to a requirement that relevant per capita, per dollar, and per spatial area impact metrics are within a globally threshold that does not undermine the sustainability of human life. One relevant global threshold, for example, is an annual per capita greenhouse gases emissions limit calculated globally as the maximum threshold to avoid unprecedented disasters due to climate change.

Life cycle sustainability assessment can be used to conduct baseline analysis, for example, on the natural resource extraction, energy, and impact intensity of materials and energy use in the current best practice of construction, operation, and decommissioning of infrastructure systems and other physical elements that affect the urban form in different ways. It can also be used to appraise future changes focusing on the life cycle performance of technical and nontechnical changes that affect the amount and type of material and energy utilization at the different stages of the life cycle of the urban infrastructure system. The appraisal process should be informed by social and economic criteria embedded in screening and prioritization tools. Integrated life cycle sustainability assessment covering the social feasibility, economic feasibility, and environmental feasibility serves as a basis for ensuring better political feasibility in city councils around the world.

4.2. Life cycle data access and quality

The future of our cities can be better shaped by more data-driven and evidence-based participatory decision-making. The three tools that make up the life cycle sustainability assessment framework are data intensive. The critical challenges of conducting the assessment are, thus, related to access to quality data. This problem gets more serious when we account for different scenarios of future technical and nontechnical changes and try to model them. Fulfilling the data requirements involves data collection, database development, and interfacing with existing data sources. We are in an era where huge amount of data resides and continues to pile up in the public and corporate realm. Data mining, access to relevant data, and presenting the data in a digestible format are part of the challenge. In the context of life cycle sustainability assessment, data sources include primary sources from cities and secondary data sources from literature such as peer-reviewed publications, reports, and generic life cycle databases. For environmental life cycle data, commercial databases are increasing available. For example, the Swiss Ecoinvent life cycle database [35] covers many product systems including energy systems in many parts of the world and other infrastructure-related datasets. Commercial databases can be purchased as part of commercial life cycle assessment software tools or as standalone databases. Free public alternative sources of life cycle data are still under development. One such example is the US Life Cycle Inventory database [36]. For social life cycle data, Social Hotspots Database (SHDB) [37] and Product Social Impact Life Cycle Assessment (PSILCA) [38] are the two available currently. For life cycle costing, generic data on material, labor, and equipment can be found from the RSMeans database [39].

4.3. Streamlined semiquantitative life cycle evaluation

Not all cities interested in sustainability issues are necessarily capable of conducting quantitative life cycle sustainability assessment. Nor is it necessary to resort to detailed quantitative assessment all the time in all contexts. There are many decision situations that only merit streamlined semiquantitative systems of accounting for all three dimensions from a life cycle perspective. The systems allow for quick assessments at a relatively low cost. They come with the capacity to capture aspects that are inherently nonquantitative. They also result in relatively easy-to-understand outcomes digestible to nonexperts. Such score-based system can take a form of a matrix structure composed of areas of protection or concern that represent the social, environment, and economic aspects on one side and the different life cycle stages on the other. The score values can be assigned based on a mix of experience, expert knowledge, previous studies, relevant checklists, and guidelines. Decision analysts working for municipalities can use such matrices or equivalent graphic systems to structure assessment information and results together with, for example, workshop-driven stakeholder perspectives in supporting decision-making.

Once relevant and critical aspects or physical assets of cities are identified using critical streamlined semiquantitative life cycle evaluation, the need for a demanding and detailed life cycle sustainability assessment that is based on quantitative environmental life cycle assessment and life cycle costing and quantitative and qualitative social life cycle assessment can be explored depending on the utility of the potential results to the decision context and resource availability.

5. The way forward

In pulling our cities out of institutional and infrastructural lock-ins and suboptimized planning and operational setting, three areas of need are identified: need for best practice demonstrations; need for framework for global urban sustainability; and need for life cycle sustainability literacy.

5.1. Best practice demonstrations

Measurable goals supported by multidimensional aspects and associated indicators are crucial in informing the planning, development, and rehabilitation projects in urban areas. The adaptation of indicators used in existing rating systems such as LEED Neighborhood and BREAM Communities to city scale is recommended [23]. Moreover, planners and developers benefit from real-world demonstration of good practices of the process and products of (re) development of districts and cities around the world.

Two city district-level cases that can be emulated by other cities customizing to local variables are Hammarby Sjöstad and Royal Seaport in Stockholm. Hammarby Sjöstad was designed in the early 1990s and developed on an old industrial area of 150 ha (200 ha with water) over the period of 12 years since 2004. Once completed, there will be 11, 000 apartments and around 35,000 people in the district [40]. It all started with an ambitious goal of becoming "twice as good" compared to the state-of-art of construction sector of the time. The detail of this goal was part of an environmental program that was adopted by the City Council of Stockholm and handed to developers. It included specific quantitative goals such as total supplied energy per square meter not exceeding 60 kWh with electricity capped at 20 kWh. It was also mentioned that the district will be built in line with the principles of natural cycles. To close the material and energy cycle locally, the Hammarby Model was later developed where solid waste and wastewater from the district are recovered in the form of electricity, transport fuel, heating, and cooking energy for use in the district [41]. The Hammarby Model captures how different systems in the district are integrated [41].

In a detailed evaluation of Hammarby Sjöstad, it was assessed as successful overall [42]. The closing of cycles has led to the reduction in metabolic flows though it is far from making the district energy wise self-sufficient [33]. The evaluation provided some recommendations for use in future (re)development projects. One such recommendation focuses on the need for integrating environmental goals early in the planning process. In Hammarby Sjöstad, the environmental program came 4 years after planning activities started. A second recommendation captured the importance of accounting for behavioral aspects of future district residents and technological limitation as part of the same holistic vision. There have recently been concerns raised by the residents of the district regarding the significant increase in the number of young people in the area, which was more than what was accounted for in the planning and development process. One aspect of the concern is the absence of a natural meeting place for the young residents within the district limits. The need to incentivize residents to live more sustainably in addition to the technical operational improvements built as part of the physical elements of the district is stressed [42].

Building on experience of Hammarby Sjöstad, a second more ambitious project on another sit in the north-east of Stockholm is under development. The Norra Djurgårdsstaden or in English the Royal Seaport started in 2009 and is right now Europe's most comprehensive urban development project pursuing the goal of creating an environmentally friendly district with at least 12,000 apartments and 35,000 residents [43]. It aims to become a fossil fuel-free district by 2030 by deploying renewable energy. Its near-term goal is limiting its per capita emission of carbon dioxide to 1.5 ton. A closed cycle model where linkages and synergies will be highlighted is also part of the plan. Developers of Royal Seaport project agreed to a legally binding building energy performance target of 55 kWh/m^2 [43]. The literature is full of connecting the concepts of smart city and sustainable city in projects like the Royal Seaport [44–46]. The political support for the Hammarby Sjöstad by Stockholm City Council was repeated for Royal Seaport by approving the environmental and sustainability program for the project in October 2010 [47]. The environmental, social, and economic sustainability goals of the district are to be realized through eight focus areas targeting technical and behavioral aspects. These are environmentally adapted residential and commercial premises; sustainable energy systems; sustainable water and wastewater systems; sustainable transport; sustainable recovery systems; climate-adapted and green outdoor environment; sustainable lifestyles; and sustainable businesses [47].

Other discussions on examples from North America include the city of Portland's Pearl District [48, 49] and Minnesota's Winona [50].

5.2. Framework for global sustainability of cities

As no city can be truly sustainable in isolation, a set of targets and indicators (e.g., [51]) to be adopted by all cities around the world is necessary. Despite its limited ambition, the UN Sustainable Development Goals (SDGs) adopted in 2015 with 2030 targets for all countries offer such a global framework that covers cities across the world. Of the 17 SDGs, SDG 11 on Sustainable Cities and Communities has most direct connection to cities. However, other goals such as SDG 3 on Good Health and Well-being; SDG 5 on Gender Equality, SDG 6 on Clean Water and Sanitation; SDG 7 on Affordable and Clean Energy; SDG 8 on Decent Work and Economic Growth; SDG 9 on Industry, Innovation and Infrastructure, and SDG 10 on Reduced Inequality are all relevant to how we develop and manage our urban areas. The same with the SDG dedicated for dealing with climate change. In a work on the network and integrated feature of SDGs, SDG 11 is presented as connected to six other SDGs [52]. Direct and indirect connections can also be made between the SDGs and the material, energy, and water flows to and from urban areas.

SDG 11 has 10 targets and 15 indicators that together recognize housing (sustainable resilient and resource-efficient buildings), basic services, waste management, transport, and public spaces as physical assets of cities. It considers natural and cultural heritage, global GDP, life and health of people as safeguard objects. A close examination of this SDG unravels implicit and explicit performance attributes of a sustainable city such as adaptation to climate change; adequacy; affordability; accessibility; convenience; free of physical and sexual harassment; holistic approach; inclusion; increase in land use efficiency (low land use rate to population growth rate ratio); integration; mitigation of climate change; participation; protection of poor and people in vulnerable situations; reduction in disaster risks; reduction in disaster-driven deaths; reduction in number of people affected by disasters; reduction in adverse per capita environmental impacts such as fine particulate matter; resource efficiency; resilience to disasters; safety; substantial decrease of economic loss; and support for positive, economic, social, and environmental link within and to surrounding areas of urban centers. SDG 11 envisions a sustainable city that avoids disasters that damage critical infrastructure and disrupt basic services.

Cities achieve these performance attributes through integrated policies and plans at urban, regional, and national levels; urban management; risk management; and local and national disaster risk reduction strategies. Environmental issues explicitly mentioned in SDG 11 are climate change, waste, air pollution, and particulate matter. This SDG mentions sustainable, resilient, and resource-efficient building and use of local building materials only and specifically in the context of least developed countries. It is not clear why the buildings of developed countries that were designed, constructed, and operated inefficiently and unsustainably are not included.

5.3. Life cycle sustainability literacy

Sustainable urban development's accounting for circumstances of constrained resources under which they occur needs to be undertaken in ways that avoid or minimize lock-in and rebound effects. Informed decisions regarding different elements of the urban fabric should be made with the goal of creating cascading positive impacts. For example, accessible golf courts that are organically integrated within the network of built-up areas of cities serve environmental functions on top of the social well-being, associated with recreation and active life as their accessibility and affordability, potentially lead to a reduction of long-distance leisure travels by local golfers and avoids or reduces impacts attributed to hotel and other related activities.

Life cycle sustainability consideration of all relevant aspects for creating new cities and redeveloping the existing comes with the complexity of working with "complementarities and contradictions within the dimensions of urban sustainability" [9]. To this end, life cycle sustainability literacy at technical and managerial level of cities is required on, for example, the material and energy ramifications of decisions passed by city council on upstream and downstream systems.

At the technical level, the capacity building task should cover on how to work with better data and state-of-the-art methodology.

At the managerial level, there is a need for understanding that life cycle sustainability assessment offers systemic view that recognizes the indivisibility of systems and sustainability even when looking at the subsystems and the individual dimensions before bringing everything into one whole.

A platform that seeks to address inequality and climate change in cities called Future Cities Canada identified areas to work on such as evaluation and impact indicators, needs-driven data, and evidence-based decision-making [53]. All these areas can be mapped into environmental and social bottom line of life cycle sustainability to be developed as part of a literacy kit for urban decision-makers.

Evidence-based decision-making at the decision-makers' and city residents' level has the potential to push cities on a sustainable development trajectory toward better sustainability. Life cycle perspective of such evidence-driven intervention recognizes long-term and lasting impacts of current decisions requiring continued commitment. Our urban world and our planet at large benefits from a depoliticized use of the life cycle view as a golden thread of planning and development. It provides a platform for engaging stakeholders along the life cycle irrespective of ideological orientation. That will potentially drive meaningful and long-lasting changes that leverage on a broader acceptance across the political spectrum.

Capacity building and awareness raising efforts will be effective if they start on streamlined semiquantitative life cycle sustainability assessment without resorting to detailed quantitative assessment.

6. Conclusions

The material and energy flow perspective of diagnosing our cities and setting goals of transforming them into more sustainable forms captures the essence of many of the global and local environmental challenges we face while it provides insights about the economic and social sustainability of what we do with the material and energy. The dominant and narrow notion of less-is-better falls short of optimal solutions as it cultivates different levels of burden shifting and suboptimization. Conventional analysis and the actions it derives only move the problem of one environmental medium into a problem of different medium. Cities have been pushing contemporary waste problems into future problems by, for example, landfilling divertible wastes. Plastic wastes from our cities and other parts of the economy have been shipped to China for so many years to the extent that it has assumed as a nonissue locally until this year when China banned the import of plastic waste from other countries including Europe and North America. The practice has been shifting the problem geographically. When the world successfully phased of chemicals that depleted the ozone layer, some of the replacements were later found to be highly potent greenhouse gases. Solving the ozone layer depletion problem was followed by unwelcome development of negatively impacting another global problem.

The consequences of the suboptimal vicious circle of solving one problem by creating another problem or parking problems for tomorrow or pushing issues spatially can be avoided at best or reduced at least by having enough level of life cycle literacy that can inform the public-level and corporate-level decisions that seek to replace old systems with new alternatives. Moving toward a thorough and comprehensive consideration of the important parts of the life cycle of material and energy used in cities, all the environmental, social, and economic aspects associated with that use within a well-defined temporal and spatial system boundary are important. Retraining current cohorts of technical and managerial staff of our cities and educating new generation of decision-makers with the skill of crafting life cycle goals and communicating to relevant stakeholders will energize such full systems-level accounting.

It is not practical or necessary to have a detailed quantitative calculation for every part of the material and energy aspects of our urban areas. Continuing with a piecemeal approach that in the long run and from a broader perspective is not cost-effective, and a disservice to the people, the planet and broad-based profit is not an option.

Author details

Getachew Assefa

Address all correspondence to: gassefa@ucalgary.ca

University of Calgary, Alberta, Canada

References

[1] Ziari K, Pourahmad A, Fotouhi Mehrabani B, Hosseini A. Environmental sustainability in cities by biophilic city approach: A case study of Tehran. International Journal of Urban Sciences. 2018:1-31. DOI: 10.1080/12265934.2018.1425153

[2] Nikulina V, Baumann H, Simon D, Sprei F. Sustainable transport futures: Analysis of the selected methodologies supporting the planning process towards achieving goal 11 sustainable cities and communities. In: Handbook of Sustainability Science and Research. Cham: Springer; 2018. pp. 473-488

[3] Hammond GP, Jones CI. Embodied energy and carbon in construction materials. Proceedings of the Institution of Civil Engineers—Energy. 2008;**161**(2):87-98

[4] Parkin S, Sommer F, Uren S. Sustainable development: Understanding the concept and practical challenge. Proceedings of the Institution of Civil Engineers—Engineering Sustainability. 2003;**156**(1):19-26

[5] Banister D, Watson S, Wood C. Sustainable cities: Transport, energy, and urban form. Environment and Planning B: Planning and Design. 1997;**24**(1):125-143

[6] Satterthwaite D. Sustainable cities or cities that contribute to sustainable development. Urban Studies. 1997;**34**(10):1667-1691

[7] Moffatt I. Edinburgh: A sustainable city? International Journal of Sustainable Development and World Ecology. 1999;**6**(2):135-148

[8] Korhonen J, Honkasalo A, Seppälä J. Circular economy: The concept and its limitations. Ecological Economics. 2018;**143**:37-46

[9] Jenks M, Jones C. Dimensions of the Sustainable City. Vol. 2. Future City, Dordrecht: Springer Netherlands; 2010

[10] Dempsey N, Brown C, Bramley G. The key to sustainable urban development in UK cities? The influence of density on social sustainability. Progress in Planning. 2012;**77**(3): 89-141

[11] Hillman T, Janson BN, Ramaswami A. Spatial allocation of transportation greenhouse gas emissions at the city scale. Journal of Transportation Engineering—ASCE. 2011;**137**(6):416-425

[12] Dodman D. Forces driving urban greenhouse gas emissions. Current Opinion in Environmental Sustainability. 2011;**3**(3):121-125

[13] Newton P. Liveable and sustainable? Socio-technical challenges for twenty-first-century cities. Journal of Urban Technology. 2012;**19**(1):81-102

[14] Jim C. Sustainable urban greening strategies for compact cities in developing and developed economies. Urban Ecosystems. 2013;**16**(4):741-761

[15] Azizalrahman H, Hasyimi V. Towards a generic multi-criteria evaluation model for low carbon cities. Sustainable Cities and Society. 2018;**39**:275-282

[16] Liqun L, Chunxia L, Yun-guang G. Green and sustainable city will become the development objective of China's low carbon city in future. Iranian Journal of Environmental Health Science & Engineering. 2014;**12**(1):34-34

[17] Bulkeley H, Betsill M. Rethinking sustainable cities: Multilevel governance and the 'urban' politics of climate change. Environmental Politics. 2005;**14**(1):42-63

[18] Hendricks MD, Meyer MA, Gharaibeh NG, Van Zandt S, Masterson J, Cooper JT, et al. The development of a participatory assessment technique for infrastructure: Neighborhood-level monitoring towards sustainable infrastructure systems. Sustainable Cities and Society. 2018;**38**:265-274

[19] Rodrigues JPC. Local government aimed at quality of life in sustainable cities. In: Entrepreneurial, Innovative and Sustainable Ecosystems. Cham: Springer; 2018. pp. 35-53

[20] Hassan AM, Lee H. The paradox of the sustainable city: Definitions and examples. Environment, Development and Sustainability. 2015;**17**(6):1267-1285

[21] Girardet H. Creating Sustainable Cities. Vol. 2. Devon, England: Resurgence Books; 1999

[22] Rauscher RC. Directions for planning sustainable cities and neighborhoods. In: New York Neighborhoods—Addressing Sustainable City Principles. Cham: Springer; 2018. pp. 217-232

[23] Barbosa JA, Bragança L, Mateus R. New approach addressing sustainability in urban areas using sustainable city models. International Journal of Sustainable Building Technology and Urban Development. 2014;**5**(4):1-9

[24] Foliente G, Kearns A, Maheepala S, Bai X, Barnett G. Beyond triple bottom line—Sustainable cities RD&D at CSIRO. 2007

[25] Dassen T, Kunseler E, Kessenich LM. The sustainable city: An analytical-deliberative approach to assess policy in the context of sustainable urban development. Sustainable Development. 2013;**21**(3):193-205

[26] Williams K. Sustainable cities: Research and practice challenges. International Journal of Urban Sustainable Development. 2010;**1**:128-132

[27] Kennedy CA, Stewart I, Facchini A, Cersosimo I, Mele R, Chen B, et al. Energy and material flows of megacities. Proceedings of the National Academy of Sciences. 2015; **112**(19):5985-5990

[28] UN. Water, Sanitation and Hygiene. Available from: http://www.unwater.org/water-facts/water-sanitation-and-hygiene. [Accessed: October 28, 2018]

[29] United Nations, Department of Economic and Social Affairs, Population Division. World Urbanization Prospects: The 2014 Revision, Highlights (ST/ESA/SER.A/352). 2014. Available from: https://esa.un.org/unpd/wup/Publications/Files/WUP2014-Highlights.pdf [Accessed: Dec 11, 2017]

[30] Hoornweg D, Bhada-Tata P. What a Waste: A Global Review of Solid Waste Management. Washington, D.C.: World Bank; 2012

[31] Hoornweg D, Bhada-Tata P, Kennedy C. Environment: Waste production must peak this century. Nature News. 2013;**502**(7473):615

[32] IPCC. Climate change 2014: Mitigation of climate change. In: Edenhofer O, Pichs-Madruga R, Sokona Y, Farahani E, Kadner S, Seyboth K, et al., editors. Contribution of Working Group III to the Fifth Assessment Report of the Intergovernmental Panel on Climate Change. Cambridge, United Kingdom, and New York, NY, USA: Cambridge University Press; 2014

[33] Iveroth SP, Johansson S, Brandt N. The potential of the infrastructural system of Hammarby Sjöstad in Stockholm, Sweden. Energy Policy. 2013;**59**:716-726

[34] Wattenbach M, Redweik R, Lüdtke S, Kuster B, Ross L, Barker A, et al. Uncertainties in city greenhouse gas inventories. Energy Procedia. 2015;**76**:388-397

[35] Wernet G, Bauer C, Steubing B, Reinhard J, Moreno-Ruiz E, Weidema B. The ecoinvent database version 3 (part I): Overview and methodology. The International Journal of Life Cycle Assessment. 2016;**21**(9):1218-1230. Available from: http://link.springer.com/10.1007/s11367-016-1087-8 [Accessed: Aug 31, 2018]

[36] U.S. Life Cycle Inventory Database. National Renewable Energy Laboratory, 2012. 2018. Available from: https://www.lcacommons.gov/nrel/search [Accessed: Aug 31, 2018]

[37] Norris CB, Norris G, Aulisio D. Social Hotspots Database. 2013. Available from: http://socialhotspot.org

[38] Ciroth A, Eisfeldt F. PSILCA—A Product Social Impact Life Cycle Assessment Database. Database Version 1.0. Berlin: GreenDelta GmbH; 2016

[39] Waier PR, Babbitt C, Baker T, Balboni B, Bastoni RA. Rockland, Massachusetts: RSMeans, Building Construction Cost Data. 2018

[40] Iverot SP, Brandt N. The development of a sustainable urban district in Hammarby Sjöstad, Stockholm, Sweden? Environment, Development and Sustainability. 2011;**13**(6):1043-1064

[41] Iveroth SP, Vernay AL, Mulder KF, Brandt N. Implications of systems integration at the urban level: The case of Hammarby Sjöstad, Stockholm. Journal of Cleaner Production. 2013;**48**:220-231

[42] Pandis S, Brandt N. Utvärdering av Hammarby Sjöstads miljöprofilering: vilka erfarenheter ska tas med till nya stadsutvecklingsprojekt i Stockholm? (Evaluation of the Hammarby Sjostad's Environmental Profile—What Experiences should be taken to new urban development projects in Stockholm?). In: Avdelningen för industriell ekologi. Stockholm: Kungliga Tekniska högskolan; 2009

[43] Shahrokni H, Årman L, Lazarevic D, Nilsson A, Brandt N. Implementing smart urban metabolism in the Stockholm Royal Seaport: Smart city SRS. Journal of Industrial Ecology. 2015;**19**(5):917-929

[44] Garau C, Pavan VM. Evaluating urban quality: Indicators and assessment tools for smart sustainable cities. Sustainability. 2018;**10**(3):575

[45] Ibrahim M, El-Zaart A, Adams C. Smart sustainable cities roadmap: Readiness for transformation towards urban sustainability. Sustainable Cities and Society. 2018;**37**:530-540

[46] Silva BN, Khan M, Han K. Towards sustainable smart cities: A review of trends, architectures, components, and open challenges in smart cities. Sustainable Cities and Society. 2018;**38**:697-713

[47] Holmstedt L, Brandt N, Robèrt KH. Can Stockholm Royal Seaport be part of the puzzle towards global sustainability?—From local to global sustainability using the same set of criteria. Journal of Cleaner Production. 2017;**140**:72-80

[48] Hall AR. Scaling sustainable infrastructure: District design for the triple bottom line [University Honors theses]. Paper 621. Portland State University; 2018

[49] Ramiller AJ. From the Neighborhood Up!: Neighborhood Sustainability Certification Frameworks and the New Urban Politics of Scale. Geography Honors Projects. 56. Macalester College; 2018. Available form: http://digitalcommons.macalester.edu/geography_honors/56 [Accessed: Oct 28, 2018]

[50] Borsari B, Mundahl N, Morse A, Mutter P, Howard JW. Achieving sustainability in the city of Winona, Minnesota (USA): A case study. In: Lifelong Learning and Education in Healthy and Sustainable Cities. Cham: Springer; 2018. pp. 327-340

[51] Koch F, Ahmad S. How to measure progress towards an inclusive, safe, resilient and sustainable city? Reflections on applying the indicators of sustainable development goal 11 in Germany and India. In: Urban Transformations. Cham: Springer; 2018. pp. 77-90

[52] Le Blanc D. Towards integration at last? The sustainable development goals as a network of targets. Sustainable Development. 2015;**23**(3):176-187

[53] FCC. 2018. Available from: https://futurecitiescanada.ca/about [Accessed: Sep 21, 2018]

Chapter 9

Smart Homes and Sustainable Cities: The Design of a Low-Cost Solution for Comprehensive Home Automation

Hubert Kalala Mukendi and Marco Adonis

Additional information is available at the end of the chapter

http://dx.doi.org/10.5772/intechopen.78058

Abstract

The challenge for smart cities is to connect as many of its inhabitants to technology enabling solutions that improve their lives. Smart homes provide all users a means of interacting and impacting their environment. In developing economies this proves challenging and these challenges are daunting and overwhelming since system costs are always a foreboding factor. The chapter addresses these challenges by providing a low-cost solution for a home energy saving measure. It introduces an overview of enabling technologies for a smart home by considering energy management, energy saving, load management and monitoring and control of living spaces. By leveraging the application of the Internet of Things (IoT) and load management strategies, the realisation of a smart home is made possible. This chapter presents a broad overview of the design and development of a web-enabled smart home solution. Web development and control systems together form the backbone of automation for modern home automation technologies such as the Internet of Things and embedded systems. The developed web-enabled home automation incorporates elements of web developed software application and digital control systems. The web-enabled interface energy saving measure is a networked system that uses web-enabled applications for enabling energy efficiency by incorporating load management, remote power consumption, monitoring and control.

Keywords: Internet of Things, smart home, energy management, energy saving, load management, energy efficiency

1. Introduction

1.1. Smart homes in a sustainable city

As more people settle in cities, the increased urbanization requires cities to become sustainable. Sustainable cities provide access to a wide range of basic services such as affordable and clean energy. Globally electricity consumption rates are increasing year on year. One way in which sustainability can be achieved is by providing an affordable range of technologies that enable the management and reduction of electricity consumption. By decreasing the demand for more electricity cities can avoid the construction of more power plants. This can be realized by adopting standards for a wider range technologies that address the reduction in electricity demand. Investment in technology, infrastructure, innovation, sustainable development and scientific research is paramount if this goal is to be achieved. Advanced tools such as information and communication technology (ICT) are able to assist the realization of sustainable cities by enabling urban dwellings to be transformed into smart homes and smart spaces. These technologies can contribute to the global efforts to provide urban citizens access to clean and affordable energy and also help the efforts to preserve the environment.

1.2. Elements of a smart home

As an assistive technology in smart cities, home automation is able to offer a high level of comfort, sophistication, data access, environmental controls, energy management and operational efficiency. Home automation provides an avenue for users to actively participate in accessing and controlling their environments. This is achieved by delivering to users tangible means of energy saving, perimeter security, data access and control of home appliances. As an energy management tool, home automation synchronizes the technology platform with practical benefits such as the ability to reduce home electric bills through control of home

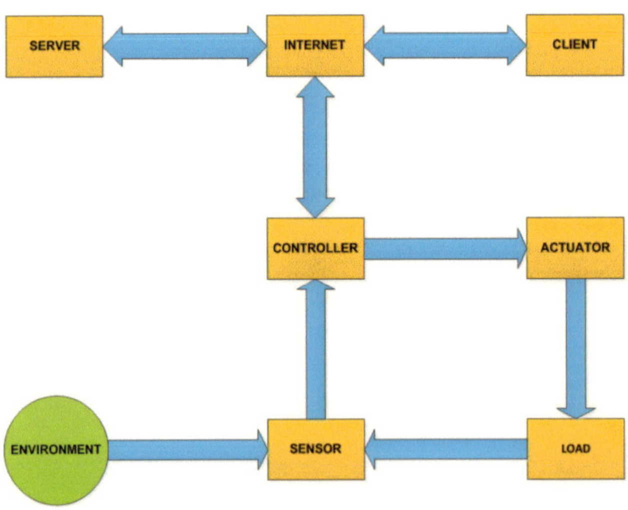

Figure 1. Home automation system elements.

electrical loads. The capability is further enhanced by the provision of remote access and resource management. This is achievable if this user desirable features include access via a web application and mobile software application. **Figure 1** indicates the typical elements of a home automation network. This represents the basic configuration encompassing sensors, actuators, controllable user loads, a main system controller and a remote user interface.

2. Literature review

2.1. Smart solutions

Modern assistive technologies should offer users smart and innovative solutions. These solutions should rely on technology to make tasks easier to manage and complete [1–3]. The rapid improvement in electronic systems such as mobile phones and home appliances are not beneficial to everyone, especially average consumers in developing countries. Energy consumption can be determined through measuring energy usage. This is accomplished using energy meters employing electrical current sensors. In electrical terms, energy consumption is the measure of how much power is used by the load side of an electrical circuit. Consumers wish to enjoy the benefits offered by smart cities such as access to information, remote sensing and automation. Facilities such as home automation can offer users some of these benefits. Recent studies have found that when users have access to their home power usage statistics, they are more readily willing to act or intervene [4, 5]. The other problem that this intervention addresses is the integrated home automation systems with customized installation where the

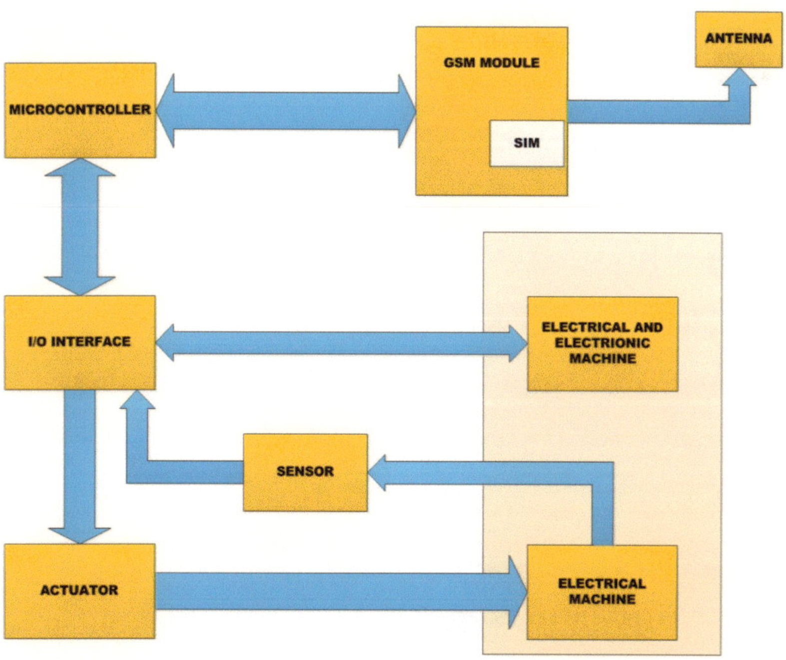

Figure 2. Mobile GSM-based home automation platform.

end users cannot make modifications for their own joy and comfort; mostly these systems are expensive and out of reach for average consumers. Inexpensive solutions are desired to address customer needs.

2.2. Home automation platforms

Recently there has been much progress made in the areas of home automation with focus on the Internet of Things (IoT) [6–8].

There have been developments in solutions presented to address home automation. The embedded controller may be provided by a host of different families of microcontrollers such as affordable variants, which include the Atmel ATmega microcontrollers [9], the Arduino [10] or Raspberry Pi [11, 12]. The propriety-based microcontrollers offer a relatively more expensive solution, which is not open-source, nor user code reconfigurable.

The communication channel has been one of the main foci. There have been solutions provided that incorporated the use of Bluetooth technology [13] as well as systems that used ZigBee communication [14]. Some of these include the use of Global System for Mobile communication or GSM network as the communication technology [15]. Such an application is presented in **Figure 2**.

Another variation on the communication channel targets the use of an Ethernet based Internet solution [16]. Such a system is depicted in **Figure 3**.

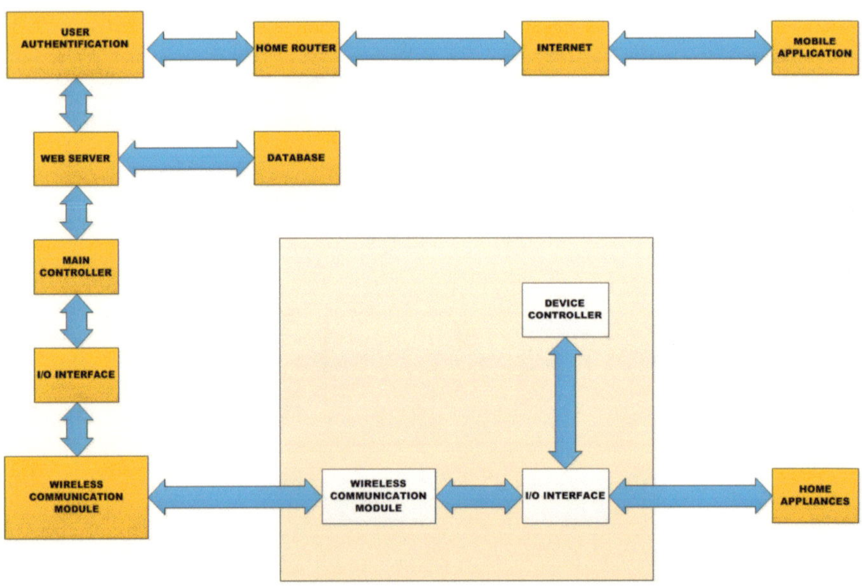

Figure 3. An Internet-Ethernet-based home automation platform.

http://dx.doi.org/10.5772/intechopen.78058

3. System design and development

3.1. Home automation solution

The system designed for this low-cost implementation uses readily available off-the-shelf components. The control algorithm is housed in an Arduino microcontroller. It is based on an open-source electronic prototyping platform and designed for ease of use. Arduino provides both the hardware components through single-board microcontrollers and also offers a software interface through its Integrated Development Environment (IDE). It is a powerful and popular embedded control solution. This makes the platform affordable for the average home user.

The solution provided here is indicated in **Figure 4**. A range of home appliances is interfaced with Arduino through its input/output terminals (I/O pins). The control and sensing actions are controlled from the microcontroller. The Arduino host the control architecture which includes the appliance database and the website server. Via the web server, users are able to access the web application on an Android device. User authentication provides added level of security allowing only authenticated users access to the system.

3.2. System algorithm

The program flowchart shown in **Figure 5** depicts the decision-making capability inherent in the system solution. The user is prompted after system initialization to authenticate themselves. If the user is not previously registered in the system database, they will not be able to proceed to the next step. If authentication passes, the user enters the home interface page.

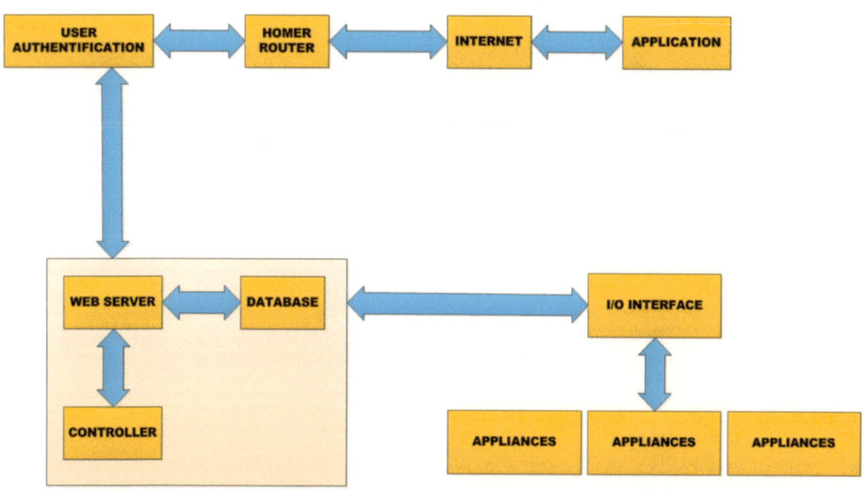

Figure 4. Implemented home automation system.

From this page the user is able to make system requests for data access to enable the appliance monitoring function or request appliance control status change. This allows connected devices such as user loads to be switched ON or OFF. The status change is indicated here in real-time.

3.3. The web interface design

A typical HTML document starts with a declaration defining the document type (see **Figure 6**), followed by the parent element of HTML document, the HTML tag. All HTML tags have an opening and a closing tag, tags are contained inside angle brackets <>. All visible contents of a web page are contained in the body tag between the opening <body> and closing </body> body tags. HTML documents are created using any text editing programs, mostly Notepad, Notepad++ and much more, and this document must be saved with a .htm or .html extensions. HTML document can be opened in any web browser, but each web browser responds differently to some HTML tags.

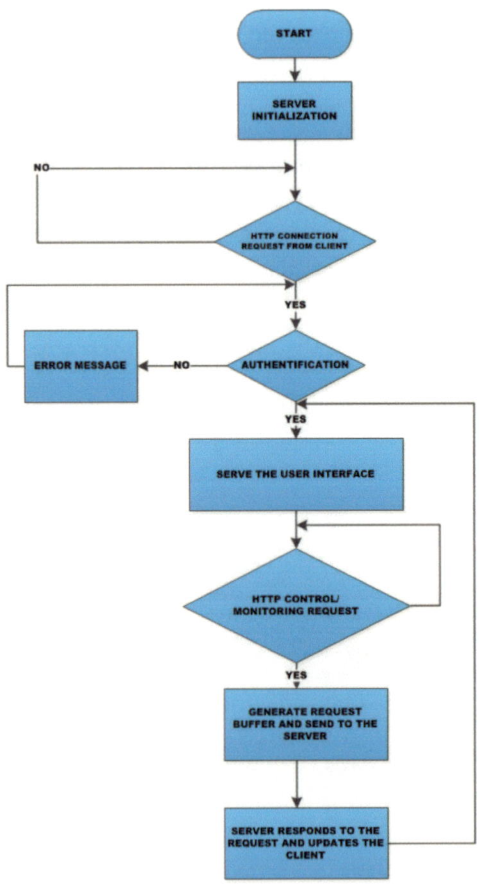

Figure 5. Program flowchart.

```
<! Doctype html>

<html>

<head>

<title>The Title Web page tile goes here</title>

</head>

<body>

<! THE VISIBLE CONTENT OF THE WEB PAGE GOES HERE -->

<h1>Big headings </h1>

<p>Paragraph</p>

</body>

</html>
```

Figure 6. HTML document structure.

The example shown below is the HTML structure, the first line in the HTML document should be the document type declaration, <! Doctype html> defines the document type and this aids the browser to correctly display the web page.

The HTML <html> tag is the root element of any HTML document, and it is the second line in the HTML document after the document type declaration. The HTML tag is closed last at the end of an HTML document. The head section in an HTML document is contained between the head opening tag <head> and the head closing tag </head>, the content of the head section is the style, title, and java scripts if any are used.

The body section in an HTML document is contained between the body opening tag <body> and the body closing tag </tag>, the body section contains the visible or information that need be displayed in the web page.

3.4. HTML design

An HTML element is made up of an opening tag <tag_name>, content and a closing tag </ tag_name>. If an HTML element does not contain any contain between the starting and end tag, it is called an empty element. The content portion of an HTML element can be another HTML element; in this case, the elements are said to be nested. All HTML documents are made of nested HTML elements, as shown below the HTML is an HTML element with opening tag <html> and closing tag </html>. The content of this element is another HTML element head which also has a starting <head> tag and a closing </head> tag. The head tag contains an element, the title tag with <title> and </title> starting and closing tags respectively. The title element contains content, Home Automation Control Panel. The HTML element also contains another element that is the body element that in turn contains an image tag. Thus the figure below is good representation HTML elements and nested elements.

The style <style> tag is used to represent the information regarding the HTML document style. The specifications pertaining to how the web browser should display the HTML document is contained in this tag. In one HTML document, it is possible to have more than one style element.

The style <style> element must be contained inside the head <head> section of the HTML document. If a "scope" attribute is specified in the style element, then that style setting will apply only to the parent and child elements that style [8].

HTML document style could also be defined in a style sheet external to the HTML document; in this case, a link tag should be used to reference the style sheet. In the example below, divBorder has defined a class with red style color. This means that any content contained within an element whose class is divBorder will have a red color.

3.5. Home automation controller elements

Figure 5 below shows a microcontroller based home automation system. The user mobile application interacts with the microcontroller via the web server using the Internet protocol. The microcontroller receives commands from the user interface and performs the required tasks based on a controlling algorithm governing the controller. The controller reads devices status and updates this data into the server for the user mobile application. Also, refer to **Figure 4** in the literature review for a typical Internet-based home automation system.

The server handles the users and ensures secure communication between the user mobile application and the controlling unit. Once a user is identified, he will then be granted access to the controlling interface (web page).

The advantages of using a microcontroller have reduced the size of circuitry, affordability, and increased flexibility. A microcontroller can be used as a substitute for other integrated circuits (IC's). It can also be easily reprogrammed to modify its functionality. The Microcontroller that was used for this project is the Arduino MEGA 2560 R3.

3.5.1. The wattmeter

The Wattmeter graphic as shown in **Figure 7**, was created using the HTML platform and Java Script. The meter was designed to measure up to 30 kW of power. The green color represents the power consumption as below 10 kW. The yellow color on the gauge represents power consumption in the range above 10–20 kW. It is an indication that the user should become aware of high power consumption. The red color represents the range above 20 kW of power consumption. Here users are acutely aware that extremely high power consumption could cause severe outage. Users should immediately switch off all non-essential appliances or high power loads.

- When the "Read Power" button is clicked, a "PowerControl()" function is called. The PowerControl function generates a random together with an HTTP "GET" request. The request buffer is made up of a "GET" method, "Power" command and a random number.
- When the web server replies to the request, the "PowerControl ()" function reads the HTTP response and changes the reading the gauge with the one contained in the response from the server.
- The random number is generated and attached to the request buffer every time the function is called to avoid caching of commands on the web browser.

3.5.2. Appliance control

A range of home appliances can be controlled via this application. These appliances are interfaced to the application controller via relays. This functionality allows users full control over appliance ON/OFF status.

Appliance control as shown in **Figure 8** below, this section of the user interface enables a person to turn ON/OFF appliances, in this case, a light bulb, a stove, a heater and a television. The system is designed in such a way that a picture representing the current state of the appliance is updated on the user interface.

Figure 7. The wattmeter user interface.

- When the Light "ON" button is clicked, the "BulbControlON()" function is called. This function generates a random number together with an HTTP "GET" request and sends it to the controller which is also the web server. The request buffer is made up of a "GET" method, "BulbON" command and a random number.
- When web server replies to the request, the "BulbControlON()" function reads the HTTP response and changes the state picture of the bulb with the one contained in the response from the server.
- When the Light "OFF" button is clicked, the "BulbControlOFF()" function is called. This function generates a random number together with an HTTP "GET" request. The request buffer is made up of a "GET" method, "BulbOFF" command and a random number.
- When web server replies to the request, the "BulbControlOFF()" function reads the HTTP response and changes the state picture of the bulb with the one contained in the response from the server.
- When the Stove "ON" button is clicked, the "StoveControlON()" function is called. This function generates a random number together with an HTTP "GET" request. The request buffer is made up of a "GET" method, "StoveON" command and a random number.
- When web server replies to the request, the "StoveControlON()" function reads the HTTP response and changes the state picture of the stove with the one contained in the response from the server.
- When the Stove "OFF" button is clicked, the "StoveControlOFF()" function is called. This function generates a random number together with an HTTP "GET" request. The request buffer is made up of a "GET" method, "StoveOFF" command and a random number.
- When web server replies to the request, the "StoveControlOFF()" function reads the HTTP response and changes the state picture of the stove with the one contained in the response from the server.
- When the Heater "ON" button is clicked, the "HeaterControlON()" function is called. This function generates a random number together with an HTTP "GET" request. The request buffer is made up of a "GET" method, "HeaterON" command and a random number.

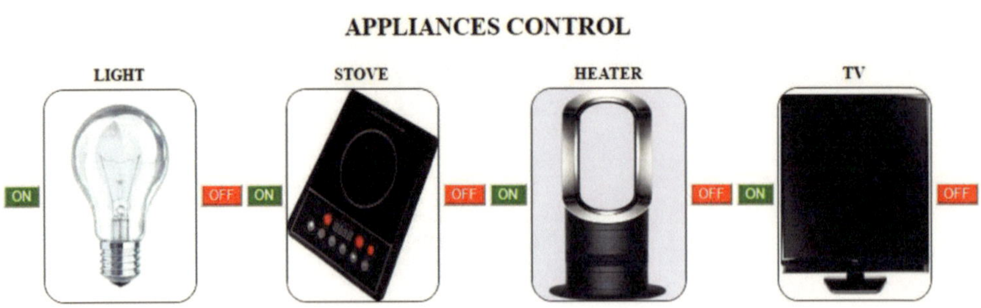

Figure 8. Home appliance control panel.

Figure 9. Security camera position and exterior light brightness control panel.

- When web server replies to the request, the "HeaterControlON()" function reads the HTTP response and changes the state picture of the heater with the one contained in the response from the server.
- When the Heater "OFF" button is clicked, the "HeaterControlOFF()" function is called. This function generates a random number together with an HTTP "GET" request. The request buffer is made up of a "GET" method, "HeaterOFF" command and a random number.
- When web server replies to the request, the "HeaterControlOFF()" function reads the HTTP response and changes the state picture of the heater with the one contained in the response from the server.
- When the TV "ON" button is clicked, the "TvControlON()" function is called. This function generates a random number together with an HTTP "GET" request. The request buffer is made up of a "GET" method, "TvON" command and a random number.
- When web server replies to the request, the "TvControlON()" function reads the HTTP response and changes the state picture of the tv with the one contained in the response from the server.
- When the TV "OFF" button is clicked, the "TvControlOFF()" function is called. This function generates a random number together with an HTTP "GET" request. The request buffer is made up of a "GET" method, "TvOFF" command and a random number.
- When web server replies to the request, the "TvControlOFF()" function reads the HTTP response and changes the state picture of the tv with the one contained in the response from the server.

3.5.3. Camera position and light brightness control

This section of the user interface as shown in **Figure 9** presents the user with means to adjust the position of a camera as well as adjust the brightness of a light. The adjustment is accomplished through two position sliders.

The camera position slider ranges from 0 to 100 with 0 representing the leftmost position of the camera or 0° position and 100 representing the rightmost position of the camera or 100° position. These positions represent the wide range of movement possible for the camera viewing postion. A camera placed on a flat wall would have 180° viewing angle. Similarly a camera placed in a corner position would either have a 90 or 270° viewing angle. See **Figure 10** for illustration of camera position and possible viewing angles.

The light brightness position slider ranges from 0 to 100% with 0% representing the light OFF and 100% representing the light ON or bright.

When the camera position or light brightness slider thumb is moved, the "PositionBrightnessControl (DeviceName, Value)" function is called. This function generates a random number together with an HTTP "GET" request. The request buffer is made up of a "GET" method, "DeviceName" command, "Value" parameter containing the desired position or brightness depending on the DeviceName and finally a random number.

This section of the user interface does not receive any response from the web server, it updates the current slider's value as it changes on the inner HTML document.

3.5.4. Intrusion detection

The intrusion detection section of the user interface as shown in **Figure 11** consists of a "CheckEntrance" button, the entrance names and the entrance status fields.

When the "CheckEntrance" button is clicked, the "IntrusionDection()" function is called. This function generates a random number together with an HTTP "GET" request. The request buffer is made up of a "GET" method, "Intrusion" command and a random number.

When web server replies to the request, the "IntrusionDection()" function reads the HTTP response and changes every entrance status field with the current state of the entrance contained in the response from the server.

Finally, the complete user interface is a web page that can be accessed via a web browser or using the android web-enabled mobile application created using App Inventor. The android app will be discussed in the next section.

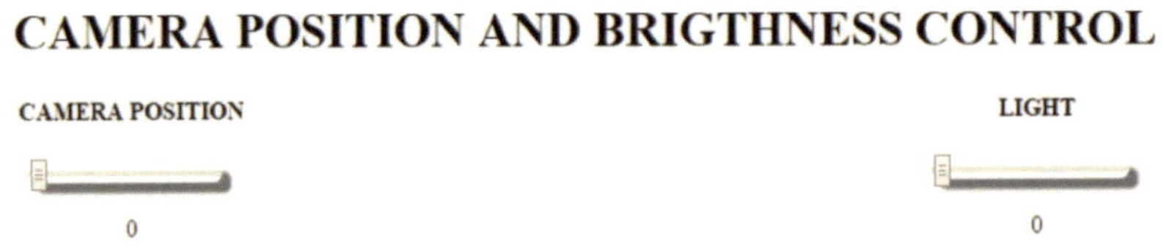

Figure 10. Camera position and possible viewing angles.

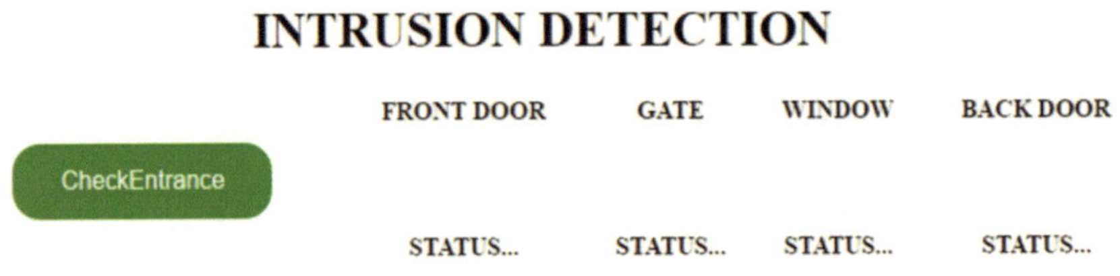

Figure 11. Security camera position and exterior light brightness control panel.

http://dx.doi.org/10.5772/intechopen.78058

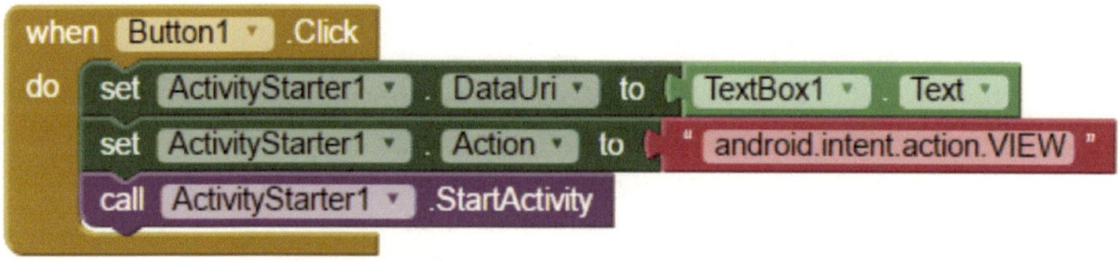

Figure 12. Designer blocks for Android app.

3.5.5. Android application design

To design the Android the web interface, the open-source Android application, APP Inventor, is used to create a simple android application that browses to a web user interface.

This App presents the user with an input field where the user can enter the IP address or domain of the web interface server of the home automation system controller. There is also a clickable button labeled "Click Here." Once this button is clicked, the app requests the web interface from the server.

The app was designed using the layout components, that is the vertical components who's width was set to fit the parent (fit the screen).

Text elements from the user interface were used as text input field to enter the desired web server IP address. The name describing the designer and the app itself were designed using the label component also from the user interface section.

The user clickable button component also found in the user interface section was used, its background color, size, and text were changed in the component editing section.

The web component responsible for loading the web page is found under activity starter section and it is an invisible component, meaning it is not seen in the designer phase of the app.

Once satisfied with the design and user interface of the app as shown in **Figure 12**, it is now time to move to the block section where blocks representing component's functionalities are grouped to create the desired function.

The user must enter the IP address assigned to the system to be controlled or its domain name, then the user must press the "CLICK HERE" button. As shown in **Figure 7** below, in this case, once the "CLICK HERE" button is clicked, an activity starter blocks are called.

First the "set ActivityStarter1.DataUri to" block is called and to it is attached the "TextBox1. Text."

Secondly the "set ActivityStarter1.Action to" block is called and to it is attached the "Text."

Finally the "call ActivityStarter1.StartActivity" block is called and the App will load the web page specified in the text field if it exists and the phone has Internet connectivity, else the app will return an error.

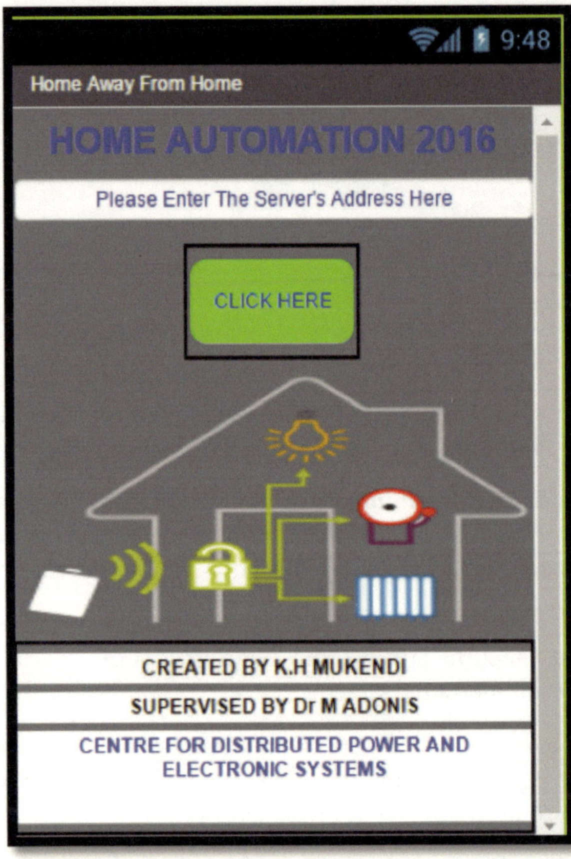

Figure 13. Android application interface panel.

3.5.6. Graphic user interface

The Android mobile user interface landing page for user authentication is shown in **Figure 13**. User authentication is applied to all users of the web app. The web server checks the credentials entered by the user and compares them with the information on its database, if the username and the associated password correspond to the ones in the database then access is granted to the user, else the user is requested to try again. When a user clicks a certain button, or requests that the status of a certain device is change, a corresponding function is called which generates a specific request buffer to send to the server. The server interprets the request buffer and controls the devices accordingly while updating the web client on the current state on the devices.

After successful user authentication, the user is directed to the home automation control interface. The fully completed and integrated user web interface monitoring and control panel is depicted in **Figure 14**. This interface houses all the various elements that encompass the total controller capability.

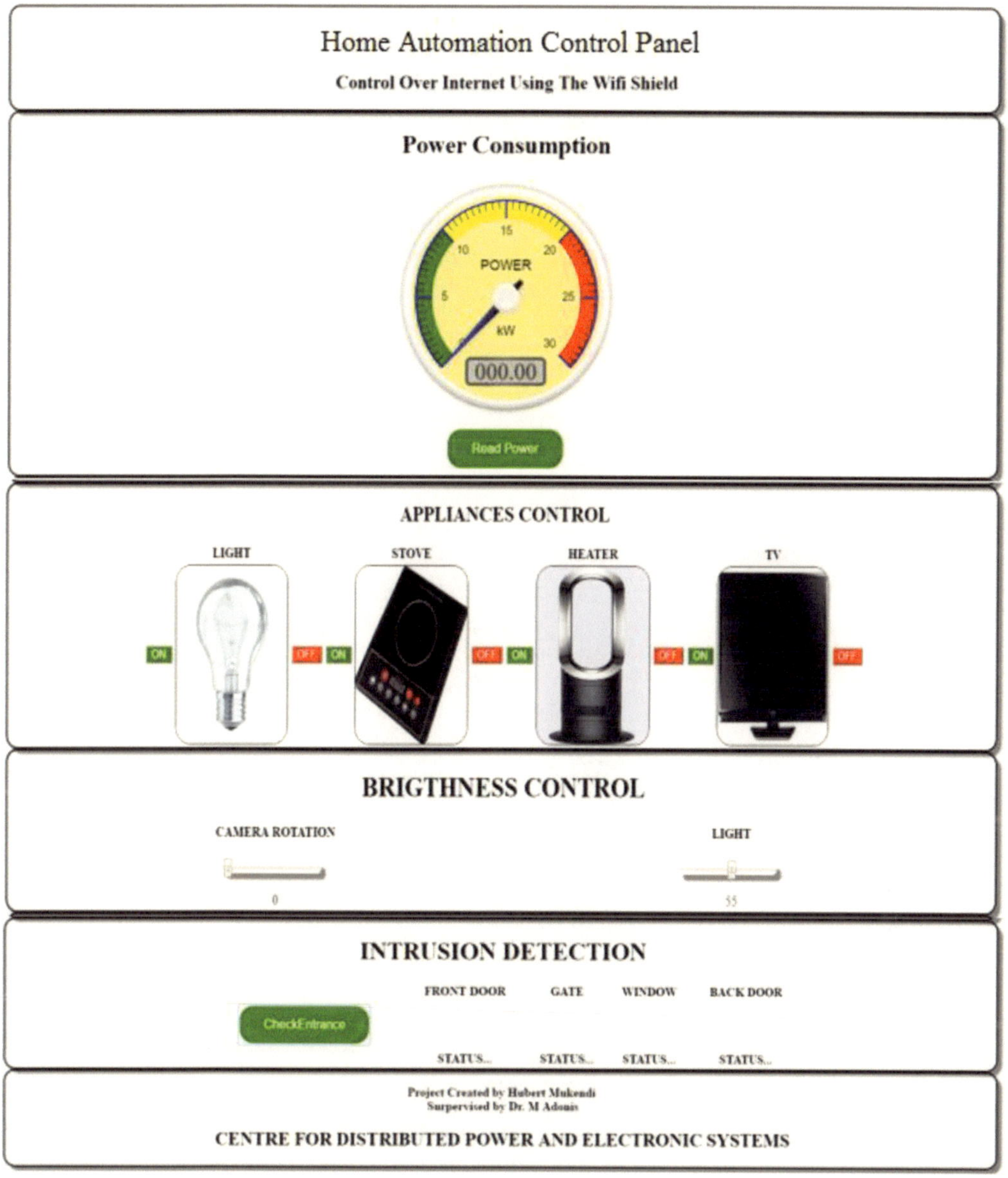

Figure 14. Complete user interface and control application.

4. Conclusion

Home automation represents a big shift in user data access and system control. As an assistive technology for smart cities, it allows users the autonomy and flexibility to interact with their immediate environment. The solution presented, required the use of the ubiquitous and low cost Arduino microcontroller. This unit is a powerful and versatile embedded platform with which to structure an intervention as described here. The implementation provided here showcases a low-cost solution for home automation. It is an effective tool for all home inhabitants to exercise control and influence over their monthly electric bills. With the application of this affordable and functional user interface, a home can easily be transformed into a smart home. It is also a solution to transform any liveable space into a smart space by offering access to measured data and allowing users regulation of the electrical loads. This comprehensive home automation solution further provides lighting dimming control and access to perimeter security through features such as security camera control and intruder detection alerts.

Author details

Hubert Kalala Mukendi* and Marco Adonis

*Address all correspondence to: mukendikalalahubert@gmail.com

Cape Peninsula University of Technology, Cape Town, South Africa

References

[1] Royston S. Smart energy technologies in everyday life: Smart Utopia? Technology Analysis & Strategic Management. 2014;**26**(10):1242-1247. ISSN 0953-7325 (Informa UK Limited). DOI: 10.1080/09537325.2014.975789

[2] Karpov AA, Lale A, Ronzhin AL. Multimodal assistive systems for a smart living environment. SPIIRAS Proceedings. 2014;**4**(19):48. ISSN 2078-9599 (SPIIRAS). DOI: 10.15622/sp.19.3

[3] Shahrestani S. Assistive IoT: Enhancing Human Experiences. Internet of Things and Smart Environments. Springer International Publishing. 2017. pp. 11-35. DOI: 10.1007/978-3-319-60164-9_2

[4] Vastardis N, Kampouridis M, Yang K. A user behaviour-driven smart-home gateway for energy management. Journal of Ambient Intelligence and Smart Environments. 2016;**8**(6):583-602. ISSN: 1876-1372 (IOS Press). DOI: 10.3233/ais-160403

[5] Liska M, Ivanic M, Volcko V, Janiga P. Research on smart home energy management system. In: 2015 16th International Scientific Conference on Electric Power Engineering (EPE). Kouty nad Desnou, Czech Republic: IEEE; 2015. DOI: 10.1109/epe.2015.7161102

[6] Kishore P, Veeramanikandasamy T, Sambath K, Veerakumar S. Internet of things based low-cost real-time home automation and smart security system. International Journal of Advanced Research in Computer and Communication Engineering. 2017;**6**(4):505-509. ISSN 2278-1021 (Tejass Publishers) DOI: 10.17148/ijarcce.2017.6497

[7] Mao X, Li K, Zhang Z, Liang J. Design and implementation of a new smart home control system based on internet of things. In: 2017 International Smart Cities Conference (ISC2). Wuxi, China: IEEE; 2017. DOI: 10.1109/isc2.2017.8090790

[8] Vimal N, Mayank P, Dipesh T, Sanket P, Yogita M. A review: Internet of things(IoT) based smart home automation. International Journal of Recent Trends in Engineering and Research. 2017;**3**(3):231-236. ISSN 2455-1457. DOI: 10.23883/ijrter.2017.3072.gbqb7

[9] Malarvizhi C, Kalaipoonguzhali V, Anitha J. Microcontroller ATmega 328Pand GSM based advanced home security system. International Journal of Smart Home. 2017;**11**(6): 11-20. ISSN 1975-4094 (Science and Engineering Research Support Society). DOI: 10.14257/ijsh.2017.11.6.02

[10] Visalatchi S, Sandeep K. Automated smart metering. Smart energy metering and power theft control using Arduino & GSM. In: 2017 2nd International Conference for Convergence in Technology (I2CT). Mumbai, India: IEEE; 2017. DOI: 10.1109/i2ct.2017.8226251

[11] Patchava V, Kandala HB, Babu PR. A smart home automation technique with raspberry Pi using IoT'. In: 2015 International Conference on Smart Sensors and Systems (IC-SSS). Bangalore, India: IEEE; 2015. DOI: 10.1109/smartsens.2015.7873584

[12] Smita M, Ravi KB. Internet of things: Smart home automation system using raspberry Pi. International Journal of Science and Research (IJSR). 2017;**6**(1):901-905. ISSN 2319-7064. DOI: 10.21275/art20164204

[13] Asadullah M, Ullah K. Smart home automation system using Bluetooth technology. In: 2017 International Conference on Innovations in Electrical Engineering and Computational Technologies (ICIEECT). Karachi, Pakistan: IEEE; 2017. DOI: 10.1109/icieect.2017.7916544

[14] Yi X-J, Zhou M, Liu J. Design of smart home control system by Internet of Things based on ZigBee. In: 2016 IEEE 11th Conference on Industrial Electronics and Applications (ICIEA). Hefei, China: IEEE; 2016. DOI: 10.1109/iciea.2016.7603564

[15] Das H, Saikia LC. GSM enabled smart energy meter and automation of home appliances. In: 2015 International Conference on Energy, Power and Environment: Towards Sustainable Growth (ICEPE). IEEE; 2015. DOI: 10.1109/epetsg.2015.7510071

[16] Wenbo Y, Quanyu W, Zhenwei G. Smart home implementation based on Internet and WiFi technology. In: 2015 34th Chinese Control Conference (CCC). IEEE; 2015. DOI: 10.1109/chicc.2015.7261075